HEAT PIPES

THE PERGAMON TEXTBOOK INSPECTION COPY SERVICE

An inspection copy of any book published in the Pergamon International Library will gladly be sent to academic staff without obligation for their consideration for course adoption or recommendation. Copies may be retained for a period of 60 days from receipt and returned if not suitable. When a particular title is adopted or recommended for adoption for class use and the recommendation results in a sale of 12 or more copies, the inspection copy may be retained with our compliments. The Publishers will be pleased to receive suggestions for revised editions and new titles to be published in this important International Library.

OTHER IMPORTANT PERGAMON TITLES OF INTEREST

AFGAN   Heat & Mass Transfer in Boundary Layers

EDE   An Introduction to Heat Transfer
                    Principles and Calculations

GRAY & MULLER   Engineering Calculations in Radiative
                    Heat Transfer

KUTATELADZE   Problems of Heat Transfer and Hydraulics
                    of Two-Phase Media

ZARIC   Heat & Mass Transfer in Flows with
                    Separated Regions

WHITAKER   Elementary Heat Transfer Analysis

WHITAKER   Fundamental Principles of Heat Transfer

PROGRESS IN HEAT AND MASS TRANSFER

GRIGULL & HAHNE   Volume 1

GOLDSTEIN et al   Volume 2
                    Eckert Presentation Volume

ECKERT & IRVINE   Volume 3
                    Heat Transfer Reviews, 1953-1969

MARTYNENKO   Volume 4
                    Luikov Presentation Volume

SCHOWALTER   Volume 5
                    Heat and Mass Transfer in Rheologically
                    Complex Fluids

HETSRONI et al   Volume 6
                    Proceedings of the International
                    Symposium on Two-Phase Systems

DWYER   Volume 7
                    Heat Transfer in liquid Metals

HEAT PIPES
Second Edition

P. Dunn

Department of Engineering and Cybernetics
University of Reading, England

and

D. A. Reay

International Research and Development Co. Ltd.,
Newcastle-Upon-Tyne, England

PERGAMON PRESS

OXFORD · NEW YORK · TORONTO · SYDNEY
PARIS · FRANKFURT

U.K.	Pergamon Press Ltd., Headington Hill Hall, Oxford OX3 0BW, England
U.S.A.	Pergamon Press Inc., Maxwell House, Fairview Park, Elmsford, New York 10523, U.S.A.
CANADA	Pergamon of Canada Ltd., 75 The East Mall, Toronto, Ontario, Canada
AUSTRALIA	Pergamon Press (Aust.) Pty. Ltd., 19a Boundary Street, Rushcutters Bay, N.S.W. 2011, Australia
FRANCE	Pergamon Press SARL, 24 rue des Ecoles, 75240 Paris, Cedex 05, France
FEDERAL REPUBLIC OF GERMANY	Pergamon Press GmbH, 6242 Kronberg/Taunus, Pferdstrasse 1, Federal Republic of Germany

First edition 1976

Reprinted with Revisions 1976

Second Edition 1978

British Library Cataloguing in Publication Data

Dunn, Peter, b. 1927

Heat pipes. - 2nd ed.
1. Heat pipes
I. Title II. Reay, David Anthony
621.4'025   TJ264   77-30268
ISBN 0-08-022127-0
ISBN 0-08-022128-9 Pbk

In order to make this volume available as economically and as rapidly as possible the authors' typescript has been reproduced in its original form. This method unfortunately has its typographical limitations but it is hoped that they in no way distract the reader.

Printed and bound in Great Britain by A. Wheaton & Co., Ltd., Exeter

Contents

Preface to Second Edition

Since the revised edition was published in 1976, with an Appendix summarizing
the proceedings of the 2nd International Heat Pipe Conference held earlier in
that year, we have been given the opportunity to prepare a second edition of
this book. This has enabled us to incorporate the data from this conference
in the appropriate chapters, and has allowed us to add further data, both from
our own and other research programmes.

Significant additions have been made to Chapters 2, 3, 5 and 7, and in some
of the applications of the heat pipe described in Chapter 7, the text has been
totally rewritten to ensure that topical and important applications are emphasized. Thus readers of the original edition, as well as new readers, will find
much additional data and comment. In conjunction with the rewriting of Chapter
7, the Bibliography on Heat Pipe Applications (Appendix 6) has been updated.

Probably one of the most interesting developments in the heat pipe field during the last two years has been the increasing emphasis on the use of wicks
to enhance the performance of thermal syphons. The 'gravity-assisted' heat
pipe, allocated a full session at the 2nd International Heat Pipe Conference,
and of particular importance in heat recovery units of the type described in
Chapter 7, is the subject of increasing development effort.

We hope that this edition proves of interest and value to the reader.

June 1977   P.D. Dunn
                                                       D.A. Reay

Preface

Following the publication by G.M. Grover et al of the paper entitled "Structures
of Very High Thermal Conductance" in 1964, interest in the heat pipe has grown
considerably. There is now a very extensive amount of literature on the
subject and the heat pipe has become recognised as an important development in
heat transfer technology.

This book is intended to provide the background required by those wishing to
use or to design heat pipes. The development of the heat pipe is discussed
and a wide range of applications described.

The presentation emphasises the simple physical principles underlying heat pipe
operation in order to provide an understanding of the processes involved.
Where necessary a summary of the basic physics is included for those who may
not be familiar with these particular topics.

Full design and manufacturing procedures are given and extensive data provided
in Appendix form for the designer.

The book should also be of use to those intending to carry out research in the
field.

The authors wish to thank those who assisted in providing material for this
book. They are especially grateful to Pru Leach for typing the major part of
the manuscript and for making many valuable suggestions concerning presentation
of the data.

Preface to Revised
Edition

Since publication of the first edition of this book in 1975, there has been a
growing interest in the study of heat pipes in Universities and Technical
Colleges. The reasons for this are not too difficult to identify: the heat
pipe can, even in its simplest form, provide a unique medium for the study of
several aspects of fluid dynamics and heat transfer, and it is growing in significance as a tool for use by the practicing engineer or physicist in applications ranging from heat recovery to precise control of laboratory experiments.

What other comparatively simple device can be used in one of its several forms
to illustrate and study the phenomena listed below ?

     Flow in porous media
     Liquid/vapour interactions
     Boiling phenomena
     Gaseous diffusion
     Condensation phenomena
     Wetting properties of materials
     Materials compatibility
     Incompressible and compressible flows
     Dynamic effects in combined heat and mass transfer
     Osmosis
     Electrokinetic phenomena
     Vapour lift and bubble pumps
     Diode behaviour, etc.

These phenomena may be examined at any selected temperature, using working
fluids ranging from liquid helium to liquid metals, depending on the degree of
sophistication of equipment available to the experimenter.

While the theory and performance of many forms of heat pipe are adequately
documented, and prediction of performance using the flow and heat transfer
equations is routine, many areas of work remain for further investigation.
Thus topics may be selected for undergraduate and advanced studies, with the
knowledge that the contribution made by the research worker is potentially
useful in later applications of the heat pipe. Our correspondence also suggests
that schools are becoming interested in the heat pipe as a piece of instructive

apparatus in physics.

Because of the growing interest in heat pipes in the field of engineering and science education, it has been decided to publish a low priced student edition of "Heat Pipes" in soft cover binding.

This has afforded us the opportunity to incorporate corrections and also to update the text. The 2nd International Heat Pipe Conference was held in Bologna from March 31st to April 2nd 1976, and a review of selected papers presented at this conference is the subject of an additional Appendix, which, when read in conjunction with the main text, extends several aspects of heat pipe theory, design, performance and application, ensuring that this public- ation remains abreast of the 'state of the art'.

P.D. Dunn
D.A. Reay
June 1976.

Introduction

The Heat Pipe and the Thermal Syphon

The heat pipe is a device of very high thermal conductance. The idea of the heat pipe was first suggested by R.S. Gaugler (1) in 1942. It was not, however, until its independent invention by G.M. Grover (2, 3) in the early 1960s that the remarkable properties of the heat pipe became appreciated and serious development work took place.

The heat pipe is similar in some respects to the thermal syphon and it is helpful to describe the operation of the latter before discussing the heat pipe. The thermal syphon is shown in Fig. 1a. A small quantity of water is placed in a tube from which the air is then evacuated and the tube sealed. The lower end of the tube is heated causing the liquid to vaporize and the vapour to move to the cold end of the tube where it is condensed. The condensate is returned to the hot end by gravity. Since the latent heat of evaporation is large, considerable quantities of heat can be transported with a very small temperature difference from end to end. Thus the structure will have a high effective thermal conductance. The thermal syphon has been used for many years and various working fluids have been employed. One limitation of the basic thermal syphon is that in order for the condensate to be returned to the evaporator region by gravitational force, the latter must be situated at the lowest point.

The heat pipe is similar in construction to the thermal syphon but in this case a wick, constructed for example from a few layers of fine gauze, is fixed to the inside surface and capillary forces return the condensate to the evaporator. (See Fig. 1b). In the heat pipe the evaporator position is not restricted and it may be used in any orientation. If, of course, the heat pipe evaporator happens to be in the lowest position gravitational forces will assist the capillary forces. The term 'heat pipe' is also used to describe high thermal conductance devices in which the condensate return is achieved by other means, for example centripetal force.

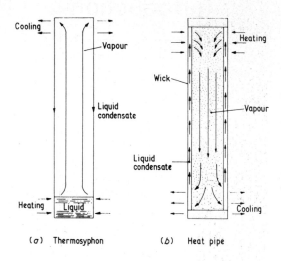

Fig. 1   The heat pipe and thermal syphon.

Several methods of condensate return are listed in Table 1.

TABLE 1   METHODS OF CONDENSATE RETURN.

Gravity	Thermal syphon
Capillary force	Standard heat pipe
Centripetal force	Rotating heat pipe
Electrostatic volume forces	Electrohydrodynamic heat pipe
Magnetic volume forces	Magnetohydrodynamic heat pipe
Osmotic forces	Osmotic heat pipe

The Heat Pipe. Construction, Performance and Properties

The main regions of the heat pipe are shown in Fig. 2.
In the longitudinal direction (see Fig. 2a), the heat pipe is made up of an
evaporator section and a condenser section. Should external geometrical
requirements make this necessary a further, adiabatic, section can be included
to separate the evaporator and condenser. The cross-section of the heat pipe,
Fig. 2b, consists of the container wall, the wick structure and the vapour
space.

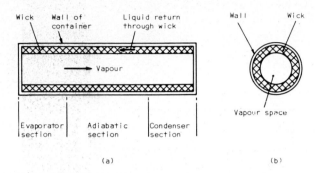

Fig. 2 The main regions of the heat pipe.

The performance of a heat pipe is often expressed in terms of 'equivalent thermal conductivity'. A tubular heat pipe of the type illustrated in Fig. 2, using water as the working fluid and operated at $150^{\circ}C$ would have a thermal conductivity several hundred times that of copper. The power handling capability of a heat pipe can be very high; pipes using lithium as the working fluid at a temperature of $1500^{\circ}C$ will carry an axial flux of 10 - 20 kW/cm^2. By suitable choice of working fluid and container materials it is possible to construct heat pipes for use at temperatures ranging from 4 K to 2300 K.

For many applications the cylindrical geometry heat pipe is suitable but other geometries can be adopted to meet special requirements.

The high thermal conductance of the heat pipe has already been mentioned; this is not the sole characteristic of the heat pipe.

The heat pipe is characterised by:

(i)   Very high effective thermal conductance.

(ii)   The ability to act as a thermal flux transformer. This
        is illustrated in Fig. 3.

(iii)   An isothermal surface of low thermal impedance. The conden-
        ser surface of a heat pipe will tend to operate at uniform
        temperature. If a local heat load is applied, more vapour
        will condense at this point, tending to maintain the temp-

erature at the original level.

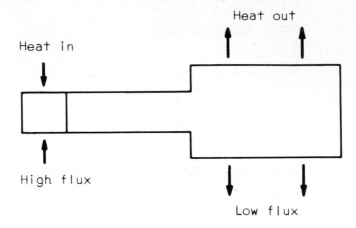

Fig. 3   The heat pipe as a thermal flux transformer.

Special forms of heat pipe can be designed having the following characteristics:

(iv) Variable thermal impedance or VCHP.
      A form of the heat pipe, known as the gas buffered heat
      pipe, will maintain the heat source temperature at an
      almost constant level over a wide range of heat input.
      This may be achieved by maintaining a constant pressure in
      the heat pipe but at the same time varying the condensing
      area in accordance with the change in thermal input. A
      convenient method of achieving this variation of condensing
      area is that of 'gas buffering'. The heat pipe is connected
      to a reservoir having a volume much larger than that of the
      heat pipe. The reservoir is filled with an inert gas which
      is arranged to have a pressure corresponding to the satur-
      ation vapour pressure of the fluid in the heat pipe. In
      normal operation a heat pipe vapour will tend to pump the
      inert gas back into the reservoir and the gas-vapour inter-
      face will be situated at some point along the condenser
      surface. The operation of the gas buffer is as follows.

      Assume that the heat pipe is initially operating under

steady state conditions. Now let the heat input increase
by a small increment. The saturation vapour temperature
will increase and with it the vapour pressure. Vapour
pressure increases very rapidly for very small increases
in temperature, for example the vapour pressure of sodium
at $800^{\circ}C$ varies as the tenth power of the temperature.
The small increase in vapour pressure will cause the inert
gas interface to recede, thus exposing more condensing
surface. Since the reservoir volume has been arranged to
be large compared to the heat pipe volume, a small change
in pressure will give a significant movement of the gas
interface. Gas buffering is not limited to small changes
in heat flux but can accommodate considerable heat flux
changes.

It should be appreciated that the temperature which is
controlled in the more simple gas buffered heat pipes, as
in other heat pipes, is that of the vapour in the pipe.
Normal thermal drops will occur when heat passes through
the wall of the evaporating surface and also through the
wall of the condensing surface.

A further improvement is the use of an active feedback
. loop. The gas pressure in the reservoir is varied by means
of an electrical heater which is controlled by a temperature
sensing element placed in the heat source.

(v)   Thermal diodes and switches.
       The former permit heat to flow in one direction only,
       whilst thermal switches enable the pipe to be switched off
       and on.

The Development of the Heat Pipe

Initially Grover was interested in the development of high temperature heat
pipes, employing liquid metal working fluids, and suitable for supplying heat
to the emitters of thermionic electrical generators and of removing heat from

the collectors of these devices. This application is described in more detail
in Chapter 7. Shortly after Grover's publication (3), work was started on
liquid metal heat pipes by Dunn at Harwell and Neu and Busse at Ispra where
both establishments were developing nuclear powered thermionic generators.
Interest in the heat pipe concept developed rapidly both for space and terrestrial applications. Work was carried out on many working fluids including
metals, water, ammonia, acetone, alcohol, nitrogen and helium.

At the same time the theory of the heat pipe became better understood; the
most important contribution to this theoretical understanding was the paper
by Cotter (4) in 1965. The manner in which heat pipe work expanded is seen
from the growth in the number of publications, following Grover's first paper
in 1964. In 1969 Cheung (5) lists 80 references; in 1970 Chisholm in his
book (6) cites 149 references, and by 1972 the NEL Heat Pipe Bibliography (7)
contained 544 references. By the end of 1976, in excess of 1000 references
to the topic were available, two International Heat Pipe Conferences had been
held, and a third was being planned for 1978, to be held in the United States.

In the first edition of this book, we stated that, with the staging of the
first International Heat Pipe Conference in Stuttgart in 1973, the heat pipe
had truely arrived. By 1977 it has become established as a most useful device
in many mundane applications, as well as retaining its more glamorous status
in spacecraft temperature control (8).

The most obvious pointer to the success of the heat pipe is the wide range of
applications where its unique properties have proved beneficial. Some of
these applications are discussed in some detail in Chapter 7, but they include
the following: electronics cooling, diecasting and injection moulding, heat
recovery and other energy conserving uses, cooking, cooling of batteries, control of manufacturing process temperatures, and as a means of transferring
heat from fluidized beds.

The Contents of This Book

Chapter 1 describes the development of the heat pipe in more detail. Chapter
2 gives an account of the theoretical basis of heat pipe operations; this is
now broadly understood though some areas exist where further work is necessary,

notably in the prediction of burn-out characteristics. Chapter 3 is concerned with the application of the theory in Chapter 2, together with other
practical considerations, to the overall design of heat pipes, and includes
a number of design examples. Chapter 4 deals with the selection of materials,
compatibility considerations including life tests and the problems of fabrication, filling and sealing. Chapter 5 describes special types of heat pipe.
Chapter 6 discusses Variable Conductance Heat Pipes and Chapter 7 describes
typical applications.

A considerable amount of data is collected together in Appendices for
references purposes.

REFERENCES

1. Gaugler, R.S. US Patent Application. Dec. 21, 1942. Published US
    Patent No. 2350348. June 6, 1944.

2. Grover, G.M. US Patent 3229759. Filed 1963.

3. Grover, G.M., Cotter, T.P., Erickson, G.F. Structures of very high
    thermal conductance. J. App. Phys. Vol. 35, p 1990, 1964.

4. Cotter, T.P. Theory of heat pipes. Los Alamos Sci. Lab. Report
    No. LA-3246-MS, 1965.

5. Cheung, H. A critical review of heat pipe theory and application
    UCRL - 50453. July 15, 1968.

6. Chisholm, D. The heat pipe. M & B Technical Library, TL/ME/2.
    Pub. Mills and Boon Ltd., London 1971.

7. McKechnie, J. The heat pipe: a list of pertinent references.
    National Engineering Laboratory, East Kilbride. Applied Heat SR.
    BIB. 2 - '72, 1972.

8. Anon. Proceedings of 2nd International Heat Pipe Conference, Bologna,
    ESA Report SP-112, Vols. 1 & 2, European Space Agency, 1976.

Historical Development

As mentioned in the Introduction, the heat pipe concept was first put forward
by R.S. Gaugler of the General Motors Corporation, Ohio, USA. In a patent
application dated December 21st, 1942, and published (1.1) as US Patent No.
2350348 on June 6th, 1944, the heat pipe is described as applied to a
refrigeration system.

According to Gaugler, the object of the invention was to "... cause
absorption of heat, or in other words, the evaporation of the liquid to a
point above the place where the condensation or the giving off of heat takes
place without expending upon the liquid any additional work to lift the
liquid to an elevation above the point at which condensation takes place." A
capillary structure was proposed as the means for returning the liquid from
the condenser to the evaporator, and Gaugler suggested that one form this
structure might take would be a sintered iron wick. The wick geometries
proposed by Gaugler are shown in Fig. 1.1. It is interesting to note the
comparatively small proportion of the tube cross-section allocated to vapour
flow in all three of his designs.

Fig. 1.1   Gaugler's proposed heat pipe wick geometries.

One form of refrigeration unit suggested by Gaugler is shown in Fig. 1.2.
The heat pipe is employed to transfer heat from the interior compartment of
the refrigerator to a pan below the compartment containing crushed ice. In
order to improve heat transfer from the heat pipe into the ice, a tubular

vapour chamber with external fins is provided, into which the heat pipe is
fitted. This also acts as a reservoir for the heat pipe working fluid.

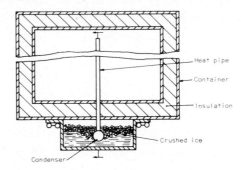

Fig. 1.2   The refrigeration unit suggested by Gaugler
in his patent published in 1944.

The heat pipe as proposed by Gaugler was not developed beyond the patent
stage, as other technology currently available at that time was applied to
solve the particular thermal problem at General Motors Corporation.

Grover's patent (1.2), filed on behalf of the United States Atomic Energy
Commission in 1963, coins the name 'heat pipe' to describe devices essentially
identical to that in the Gaugler patent. Grover, however includes a limited
theoretical analysis and presents results of experiments carried out on
stainless steel heat pipes incorporating a wire mesh wick and sodium as the
working fluid. Lithium and silver are also mentioned as working fluids.

An extensive programme was conducted on heat pipes at Los Alamos Laboratory,
New Mexico, under Grover, and preliminary results were reported in the first
publication on heat pipes (1.3). Following this the United Kingdom Atomic
Energy Laboratory at Harwell started similar work on sodium and other heat
pipes (1.4). The Harwell interest was primarily the application to nuclear
thermionic diode converters, a similar programme commenced at the Joint
Nuclear Research Centre, Ispra, in Italy under Neu and Busse. The work at
Ispra built up rapidly and the laboratory became the most active centre for
heat pipe research outside the US, (1.5, 1.6).

The work at Ispra was concerned with heat pipes for carrying heat to emitters

and for dissipating waste heat from collectors. This application
necessitated heat pipes operating in the temperature regions between about
$1600^{\circ}C$ and $1800^{\circ}C$ (for emitters) and $1000^{\circ}C$ (for collectors). At Ispra the
emphasis was on emitter heat pipes, which posed the more difficult problems
concerning reliability over extended periods of operation.

The first commercial organisation to work on heat pipes was RCA (1.7, 1.8).
Most of their early support came from US Government contracts, during the
two year period mid-1964 to mid-1966 they made heat pipes using glass, copper,
nickel, stainless steel, molybdenum and TZM molybdenum as wall materials.
Working fluids included water, caesium, sodium, lithium and bismuth. Maximum
operating temperatures of $1650^{\circ}C$ had been achieved.

Not all of the early studies on heat pipes involved high operating temperatures. Deverall and Kemme (1.9) developed a heat pipe for satellite use
incorporating water as the working fluid, and the first proposals for a
variable conductance heat pipe were made again for a satellite (1.10). (The
variable conductance heat pipe is fully described in Chapter 6).

During 1967 and 1968 several articles appeared in the scientific press, most
originating in the United States, indicating a broadening of the areas of
application of the heat pipe to electronics cooling, air conditioning, engine
cooling and others (1.11, 1.12, 1.13). These revealed developments such as
flexible and flat plate heat pipes. One point stressed was the vastly
increased thermal conductivity of the heat pipe when compared with solid
conductors such as copper, a water heat pipe with a simple wick having an
effective conductivity several hundred times that of a copper rod of similar
dimensions.

Work at Los Alamos Laboratory continued at a high level. Emphasis was still
on satellite applications, and the first flights of heat pipes took place in
1967 (1.9). In order to demonstrate that heat pipes would function normally
in a space environment, a water/stainless steel heat pipe, electrically
heated at one end, was launched into an earth orbit from Cape Kennedy on an
Atlas-Agena vehicle. When the orbit had been achieved the heat pipe was
automatically turned on and telemetry data on its performance was successfully received at 5 tracking stations in a period lasting 14 orbits. The

data suggested that the heat pipe operated successfully.

By now the theory of the heat pipe was well developed, based largely on the
work of Cotter (1.14), also working at Los Alamos. So active were
laboratories in the United States and at Ispra that in his critical review of
heat pipe theory and applications (1.15), Cheung was able to list over 80
technical papers on all aspects of heat pipe development. He was able to
show that the reliabllity of liquid metal heat pipes under long term operation (9000 hr) at elevated temperatures (1500°C) had been demonstrated.
Heat pipes capable of transferring axial fluxes of 7 kW/cm^2 had been constructed, and plans were in hand to more than double this capability. Radial heat
fluxes of up to 400 W/cm^2 had been achieved.

Cheung also referred to various forms of wick, including an arterial type
illustrated in Fig. 1.3, which was developed by Katzoff (1.16). Its operation was tested in a glass heat pipe using an alcohol as the working fluid.
The function of the artery, which has become a common feature of heat pipes
developed for satellite use, is to provide a low pressure drop path for
transporting liquid from the condenser to the evaporator, where it is
redistributed around the heat pipe circumference using a fine pore wick
provided around the heat pipe wall.

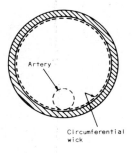

Fig. 1.3 An arterial wick form developed by Katzoff (1.16)

Following the first heat pipe test in space in 1967 (1.9) the first use of
heat pipes for satellite thermal control was on GEOS-B, launched from
Vandenburg Air Force Base in 1968 (1.17). Two heat pipes were used, located
as shown in Fig. 1.4. The heat pipes were constructed using 6061 T-6 aluminium alloy, with 120 mesh aluminium as the wick material. Freon 11 was used

as the working fluid. The purpose of the heat pipes was to minimise the
temperature differences between the various transponders in the satellite.
Based on an operating period of 145 days, the range between the maximum and
minimum transponder temperatures was considerably smaller than that in a
similar arrangement in GEOS-A, an earlier satellite not employing heat pipes.
The heat pipes operated near-isothermally and performed well over the
complete period of observation.

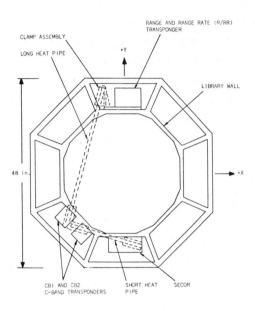

Fig. 1.4 Heat pipes used in space - the GEOS-B satellite

In 1968 Busse wrote a paper (1.18) which summarised the heat pipe activities
in Europe at that time, and it is notable that the Ispra laboratory of
Euratom was still the focal point for European activities. Other laboratories
were making contributions, however, including Brown Boveri, Karlsruhe Nuclear
Research Centre, the Institüt für Kernenergetic (IKE) Stuttgart, and Grenoble
Nuclear Research Centre. The experimental programmes at the above laboratories were performed largely on heat pipes using liquid metals as working
fluids, and centred on life tests and measurements of the maximum axial and
radial heat fluxes. Theoretical aspects of heat transport limitations were
also studied. By now we were also seeing the results of basic studies on
separate features of heat pipes, for example wick development, factors
affecting evaporator limiting heat flux, and the influence of non-condensable

gases on performance.

In Japan a limited experimental programme was conducted at the Kisha Seizo
Kaisha Company (1.19). Presenting a paper on this work in April 1968 to an
audience of air conditioning and refrigeration engineers, Nozu described an
air heater utilising a bundle of finned heat pipes. This heat pipe heat
exchanger is of considerable significance in the current energy situation, as
it can be used to recover heat from hot exhaust gases, and can be applied in
industrial and domestic air conditioning. Such heat exchangers are now
available commercially and are referred to in Chapter 7 and Appendix 9.

During 1969 the published literature on heat pipes showed that establishments
in the United Kingdom were increasingly aware of their potential including
the British Aircraft Corporation and the Royal Aircraft Establishment (RAE),
Farnborough. RAE (1.20) were evaluating heat pipes and vapour chambers for
the thermal control of satellites, and BAC had a similar interest.

It was during this year that IRD commenced work on heat pipes, initially in
the form of a survey of potential applications, followed by an experimental
programme concerned with the manufacture of flat plate and tubular heat pipes.
Some examples of these are shown in Fig. 1.5. Work was also being carried out
under Dunn at Reading University, where several members of the staff had
experience of the heat pipe activities at Harwell, described earlier. The
National Engineering Laboratory (NEL) at East Kilbride and the National Gas
Turbine Establishment (NGTE) at Pyestock also entered the field.

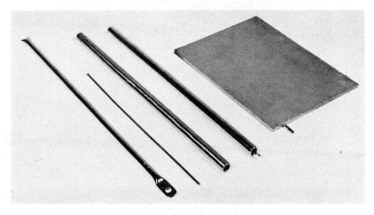

Fig. 1.5 Flat plate and tubular heat pipes (Courtesy IRD)

Interest in the USSR in heat pipes was evident from an article published in
the Russian journal 'High Temperature' (1.21) although much of the information described a summary of work published elsewhere.

1969 saw reports of further work on variable conductance heat pipes, the
principle contributions being made by Turner (1.22) at RCA and Bienert (1.23)
at Dynatherm Corporation. Theoretical analyses were carried out on VCHP's to
determine parameters such as reservoir size, and practical aspects of reservoir construction and susceptibility to external thermal effects were
considered.

A new type of heat pipe, in which the wick is omitted, was developed at this
time by NASA (1.24). The rotating heat pipe utilises centrifugal acceleration
to transfer liquid from the condenser to the evaporator, and can be used for
cooling motor rotors and turbine blade rotors. Gray (1.24) also proposed an
air conditioning unit based on the rotating heat pipe, and this is illustrated in Fig. 1.6. (The rotating heat pipe is described fully in Chapter 5.)
The rotating heat pipe does not of course suffer from the capillary pumping
limitations which occur in conventional heat pipes, and its transport
capability can be greatly superior to that of wicked pipes.

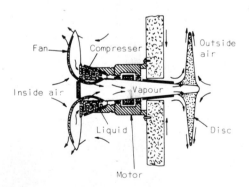

Fig. 1.6 A compact air conditioning unit based on the
wickless rotating heat pipe. (Courtesy NASA)

The application of heat pipes to electronics cooling in areas other than
satellites was beginning to receive attention. Pipes of rectangular section
were proposed by Sheppard (1.25) for cooling integrated circuit packages, and
the design, development and fabrication of a heat pipe to cool a high-power

airborne travelling wave tube (TWT) is described by Calimbas and Hulett of
the Philco-Ford Corporation (1.26).

Most of the work on heat pipes described so far has been associated with
liquid metal working fluids and, for lower temperatures, water, acetone, the
alcohols etc. With the need for cooled detectors in satellite infra-red
scanning systems, to mention but one application, cryogenic heat pipes began
to receive particular attention (1.27, 1.28). The most common working fluid
in these heat pipes was nitrogen, which was acceptable for temperature ranges
between 77 and 100°K. Liquid oxygen was also used for this temperature range.
The Rutherford High Energy Laboratory was the first organisation in the United
Kingdom to operate cryogenic heat pipes (1.29), liquid hydrogen units being
developed for cooling targets at the RHEL. Later RHEL developed a helium heat
pipe operating at 4.2°K (1.30).

By 1970 a wide variety of heat pipes were commercially available from a number
of companies in the United States. RCA, Thermo-Electron, and Noren Products
were among several firms marketing a range of 'standard' heat pipes, with the
ability to construct 'specials' for specific customer applications. During
the next few years several manufacturers were established in the United
Kingdom (see Appendix 5) and a number of companies specialising in heat pipe
heat recovery systems, based primarily on technology from the United States,
have entered what is becoming an increasingly competitive market (1.31).

The early 1970's saw a considerable growth in the application of heat pipes
to solve terrestrial heat transfer problems, in addition to the continuing
momentum in their development for spacecraft thermal control. The European
Space Agency commenced the funding of research and development work with
companies and universities in Britain and Continental Europe, with a view to
producing space-qualified heat pipes for ESA satellites. Other companies,
notably Dornier, SABCA, Aerospatiale and Marconi, put considerable effort
into development programmes, initially independent of ESA, in order to keep
abreast of the technology. As a result of this, a considerable number of
European companies can compete effectively in the field of spacecraft thermal
control using heat pipes. (While the application of heat pipes in spacecraft
may seem rather esoteric to the majority of potential users of these devices,
the 'technological fallout' has been considerable, both from this and appli-

cations in nuclear engineering, and has contributed significantly to the
design procedures, reliability, and performance improvements of the commercial
products over the past decade.)

The effort expended in developing variable conductance heat pipes (VCHPs),
initially to meet the requirements of the US space programme, has led to one
of the most significant types of heat pipe, one which is able to effect precise temperature control. The number of techniques available for control of
heat pipes is large, and some are discussed in Chapter 6. Reference 1.32
gives an excellent summary of the state-of-the-art of this technology.

One of the major engineering projects of the 1970's, the construction and
operation of the trans-Alaska oil pipeline, makes use of heat pipe technology.
As described in Chapter 7, heat pipes are used in the pipeline supports in
order to prevent melting of the permafrost, and the magnitude of the project
necessitates McDonnell Douglas Astronautics Company producing 12000 heat pipes
per month, the pipes ranging in length between 9 and 23 metres.

While much development work was concentrated on 'conventional' heat pipes, the
last two or three years have seen increasing interest in the rotating heat
pipe and in research into electrohydro-dynamics for liquid transport. The
proposed use of 'inverse' thermal syphons and an emphasis on the advantages
(and possible limitations) of gravity-assisted heat pipes (see Chapter 2)
have stood out as areas of considerable importance, and as such are given more
space in this edition of 'Heat Pipes'. The reason for the growing interest
in these topics is not too difficult to find. Heat pipes in terrestrial applications have, probably in the majority of cases, proved particularly viable
when gravity, in addition to capillary action, has aided condensate return to
the evaporator. This is seen best of all in heat pipe heat recovery units,
where slight changes in the inclination of the long heat pipes used in such
heat exchanges can, as soon as reliance on the wick alone is effected, can
cut off heat transport completely. (Such a technique is used, on purpose, to
control the performance of such units, and these heat exchangers are discussed
more fully in Chapter 7.)

REFERENCES

1.1   Gaugler, R.S. Heat transfer device. US Patent, 2350348, Appl.
         21 December, 1942. Published 6 June, 1944.

1.2   Grover, G.M. Evaporation-condensation heat transfer device. US
         Patent 3229759, Appl. 2 Dec. 1963. Published 18 January, 1966.

1.3   Grover, G.M., Cotter, T.P. and Erikson, G.F. Structure of very
         high thermal conductance. J. Appl. Phys. 1964, 35 (6), pp 1990 -
         1991.

1.4   Bainton, K.F. Experimental heat pipes. AERE-M1610, Harwell, Berks.
         Atomic Energy Research Establishment, Appl. Physics Div. June, 1965.

1.5   Grover, G.M., Bohdansky, J. and Busse, C.A. The use of a new heat
         removal system in space thermionic power supplies. EUR 2229e,
         Ispra, Italy. Euratom Joint Nuclear Research Centre, 1965.

1.6   Busse, C.A., Caron, R. and Cappelletti, C. Prototypes of heat pipe
         thermionic converters for space reactors. IEE, 1st Conf. on
         Thermionic Electrical Power Generation, London, 1965.

1.7   Leefer, B.I. Nuclear Thermionic energy converter. 20th Annual
         Power Sources Conf. Atlantic City, N.J. 24 - 26 May, 1966.
         Proceedings, pp 172 - 175, 1966.

1.8   Judge, J.F. RCA test thermal energy pipe. Missiles and Rockets,
         Feb. 1966, 18, pp 36 - 38.

1.9   Deverall, J.E., and Kemme, J.E. Satellite heat pipe. USAEC Report
         LA-3278, Contract W-7405-eng-36. Los Alamos Scientific Laboratory,
         University of California, Sept., 1970.

1.10   Wyatt, T. A controllable heat pipe experiment for the SE-4
         satellite. JHU Tech. Memo APL-SDO-1134. Johns Hopkins University,
         Appl. Physics Lab., March 1965. AD 695 433.

1.11   Feldman, K.T. and Whiting, G.H. The heat pipe and its potential-
         ities. Engrs. Dig., London 1967, 28 (3), pp 86 - 86.

1.12   Eastman, G.Y. The heat pipe. Scient. American, 1968, 218 (5),
         pp 38 - 46.

1.13   Feldman, K.T. and Whiting, G.H. Applications of the heat pipe.
         Mech. Engng., Nov. 1968, 90, pp 48 - 53, (US).

1.14   Cotter, T.P. Theory of heat pipes, USAEC Report LA-3246. Contract
         W7405-eng-36. Los Alamos Scientific Laboratory, Univeristy of
         California, 1965.

1.15   Cheung, H. A critical review of heat pipe theory and applications.
         USAEC Report UCRL-50453. Lawrence Radiation Laboratory, University
         of California, 1968.

1.16   Katzoff, S. Notes on heat pipes and vapour chambers and their
         applications to thermal control of spacecraft. USAEC Report
         SC-M-66-623. Contract AT (29-1)-789. Proceedings of Joint Atomic
         Energy Commission/Sandia Laboratories Heat Pipe Conference, Vol. 1,
         held at Albuquerque, New Mexico, 1 June, 1966, pp 69 - 89. Sandia
         Corporation, October, 1966.

1.17   Anand, D.K. Heat pipe application to a gravity gradient satellite.

Proc. of ASME Annual Aviation and Space Conf., Beverley Hills, California. 16 - 19 June 1968, pp 634 - 658.

1.18   Busse, C.A.   Heat pipe research in Europe. 2nd Int. Conf. on Thermionic Elec. Power Generation, Stresa, Italy, May 1968. Report EUR 4210, f, e; 1969, pp 461 - 475.

1.19   Nozu, S. et al.   Studies related to the heat pipe. Trans. Soc. Mech. Engrs. Japan, 1969, 35 (2), pp 392 - 401 (In Japanese).

1.20   Savage, C.J.   Heat pipes and vapour chambers for satellite thermal balance. RAE TR-69125, Royal Aircraft Establishment, Farnborough, Hants., June, 1969.

1.21   Moskvin, Ju. V. and Filinnov, Ju. N.   Heat pipes. High Temp., 1969 7 (6) pp 704 - 713.

1.22   Turner, R.C.   The constant temperature heat pipe. A unique device for the thermal control of spacecraft components. AIAA 4th Thermophys. Conf. Paper 69-632, San Francisco, June 16 - 19, 1969.

1.23   Bienert, W.   Heat pipes for temperature control. 4th Intersoc. Energy Conversion Conf., Washington D.C. Sept. 1969, pp 1033 - 1041

1.24   Gray, V.H.   The rotating heat pipe - a wickless hollow shaft for transferring high heat fluxes. ASME Paper 69-HT-19 New York, American Society of Mechanical Engineers, 1969.

1.25   Sheppard, T.D., Jr. Heat pipes and their application to thermal control in electronic equipment. Proc. National Electronic Packaging and Prodn. Conf., Anaheim, California, 11 - 13 Feb., 1969.

1.26   Calimbas, A.T. and Hulett, R.H.   An avionic heat pipe. ASME Paper 69-HT-16 New York, American Society of Mechanical Engineers, 1969.

1.27   Eggers, P.E. and Serkiz, A.W.   Development of cryogenic heat pipes. ASME 70-WA/Ener-1. American Society of Mechanical Engineers, 1970.

1.28   Joy, P. Optimum cryogenic heat pipe design. ASME Paper 70-HT/SpT-7. American Society of Mechanical Engineers, 1970.

1.29   Mortimer, A.R.   The heat pipe. Engineering Note-Nimrod/NDG/70-34. Harwell: Rutherford Laboratory, Nimrod Design Group, Oct. 1970.

1.30   Lidbury, J.A.   A helium heat pipe. Engineering Note NDG/72-11. Harwell: Rutherford Lab., Nimrod Design Group, April, 1972.

1.31   Reay, D.A. Industrial Energy Conservation: A Handbook for Engineers and Managers. Pergamon Press, Oxford, 1977.

1.32   Groll, M. and Kirkpatrick, J.P. Heat pipes for spacecraft temperature control—an assessment of the state-of-the-art. Proc. 2nd Int. Heat Pipe Conference, Bolgona, Italy; ESA Report SP112, Vol. 1, 1976.

Theory of the Heat Pipe

2.1 Introduction

In this chapter we discuss the theory of the heat pipe. In order for the heat pipe to operate the maximum capillary pumping head $(\Delta P_c)_{max}$ must be greater than the total pressure drop in the pipe. This pressure drop is made up of three components.

(a) The pressure drop ΔP_ℓ required to return the liquid from the condenser to the evaporator

(b) The pressure drop ΔP_v neccessary to cause the vapour to flow from the evaporator to the condenser

(c) The gravitational head ΔP_g which may be zero, positive or negative.

Thus

$$(\Delta P_c)_{max} \geqslant \Delta P_\ell + \Delta P_v + \Delta P_g \qquad \qquad \text{... 2.1.1}$$

If this condition is not met, the wick will dry out in the evaporator region and the pipe will not operate. The phenomenon of surface tension and the associated capillary effects are briefly summarised, and an expression obtained for capillary head. The components of pressure drop ΔP_ℓ, ΔP_v are then discussed. The latter is surprisingly complex but a simple expression is given to enable ΔP_v to be calculated approximately and the current theoretical position is outlined.

During start-up and with certain high temperature liquid metal pipes the vapour velocity may reach sonic values. In such cases compressibility effects must be taken into account. Sonic conditions set one limit to the maximum possible heat transport capability of a heat pipe. Other limits are set, at low temperatures, by viscous forces; and at increasing temperatures limits arise due to entrainment of the working fluid in the wick by the vapour stream, by insufficient capillary head and by evaporator burnout.

These limitations (2.1) on maximum axial heat transport are shown in Fig. 2.1.

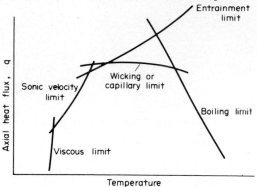

Fig. 2.1 Limitations to heat transport in
the heat pipe

It is necessary for the operating point to be chosen in the area lying below
these curves.   The actual shape of this area depends on the working fluid
and wick material and will vary appreciably for different heat pipes.   It
is shown that if vapour pressure loss and gravitational head can be
neglected the properties of the working fluid which determine maximum heat
transport can be combined to form a figure of merit M.

$$ M \; = \; \frac{\rho_\ell \sigma_\ell L}{\mu_\ell} \qquad\qquad \text{... 2.1.2} $$

where ρ_ℓ is the density of the liquid working fluid

σ_ℓ the surface tension

L   the enthalpy of vaporisation or latent heat

μ_ℓ the viscosity of the liquid working fluid

The way in which M varies with temperature for a number of working fluids is
shown in Fig. 2.2, other factors such as cost will also influence the choice
of working fluid and this topic is discussed in detail in Chapter 3.

The heat pipe has a very high thermal conductance; however, temperature
drops will occur, both radially at the evaporator and condenser and axially
down the pipe.   Formulae are given to enable these temperature differences
to be estimated.

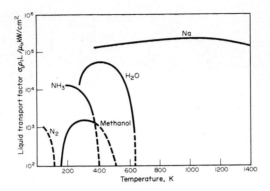

Fig. 2.2 Merit number of selected working fluids

2.2 Surface Tension and Surface Energy

2.2.1 Introduction.   Molecules in a liquid attract one another.   A
molecule in a liquid will be attracted by the other molecules around it and,
on average, will be equally attracted in all directions and hence will not
experience any resultant force.   In the case of a molecule at or near the
surface of a liquid the forces of attraction will no longer balance out and
the molecule will experience a resultant force inwards.   Because of this
effect a liquid will tend to take up a shape having minimum area, in the case
of a free falling drop in vacuum this will be a sphere.   Due to this
spontaneous tendency to contract a liquid surface behaves rather like a
rubber membrane under tension.

In order to increase the surface area work will need to be done on the
liquid.   The energy associated with this work is known as the Free Surface
Energy, and the corresponding Free Surface Energy/unit area of surface is
given the symbol σ_{ℓ}.   For example if a soap film is set up on a wire support
as in Fig. 2.3 (a) and the area is increased by moving one side a distance dx
the work done = fdx = increase in surface energy = $2\sigma_{\ell}$ldx.

The factor 2 arises since the film has two free surfaces.   Hence if T is the
force/unit length for each of the two surfaces 2Tldx = $2\sigma_{\ell}$ldx or T = σ_{ℓ}.

This force/unit length is known as the Surface Tension.   It is numerically
equal to the surface energy/unit area measured in any consistent set of units,
e.g. N/m.

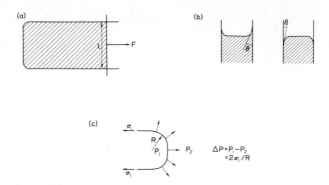

Fig. 2.3 Representation of surface tension and pressure difference across a curved surface

Values for surface tension for a number of fluids are given in Appendix 1.

Since latent heat of vaporisation L is a measure of the forces of attraction between the molecules of a liquid we might expect surface energy or surface tension σ_ℓ to be related to L.   This is found to be the case.   Solids also will have a free surface energy and in magnitude it is found to be similar to the value for the same material in the m\. 'en state.

When a liquid is in contact with a solid surface molecules in the liquid adjacent to the solid will experience forces from the molecules of the solid in addition to the forces from other molecules in the liquid.   Depending on whether these solid/liquid forces are attractive or repulsive the liquid solid surface will curve upwards or downwards, Fig. 2.3 (b).   The two best known examples of attractive and repulsive forces respectively are water and mercury.   Where the forces are attractive the liquid is said to 'wet' the solid.   The angle of contact made by the liquid surface with the solid is known as the angle of contact θ.   For wetting θ will lie between 0 and $\frac{\pi}{2}$ and for non wetting liquids $\theta > \frac{\pi}{2}$.

The condition for wetting to occur is that the total surface energy is reduced by wetting

$$\sigma_{s\ell} + \sigma_{\ell v} < \sigma_{sv}$$

where the subscripts, s, ℓ, v refer to solid, liquid and vapour phases
respectively, as shown in Fig. 2.4.

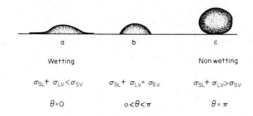

Fig. 2.4 Wetting and non-wetting contact

Wetting will not occur if

$$\sigma_{s\ell} + \sigma_{\ell v} > \sigma_{sv}, \quad \text{see Fig. 2.4 (c)}$$

The intermediate condition of partial wetting

$$\sigma_{s\ell} + \sigma_{\ell v} = \sigma_{sv} \quad \text{is shown in Fig. 2.4 (b)}$$

2.2.2 Pressure difference across a curved surface.

One of the consequences
of surface tension is that the pressure on the concave surface is less than
that on the convex surface.   This pressure difference ΔP is related to the
surface energy σ_ℓ and radius of curvature R of the surface, Fig. 2.3 (c).
This relation may be obtained as follows:

If we consider a hemispherical surface the tension forces acting round the
circumference are given by $2\pi R\sigma_\ell$ and are balanced by the pressure forces
acting across the surface $\Delta P \times \pi R^2$.   Hence $\Delta P = \dfrac{2\sigma_\ell}{R}$ ... 2.2.1

If the surface has two radii of curvature at right angles R_1, R_2, then it
can be shown

$$\Delta P = \sigma_\ell \left(\frac{1}{R_1} + \frac{1}{R_2} \right)$$

Due to this pressure difference, if a vertical tube, radius r, is placed in a liquid which wets the material of the tube, the liquid will rise in the tube to a height h above the plane surface of the liquid, Fig. 2.5.

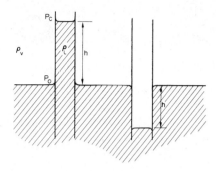

Fig. 2.5 Capillary rise in a tube

The pressure balance gives

$$\rho_\ell g h = \frac{2\sigma_\ell}{r} \cos\theta$$

where ρ_ℓ is the liquid density
and θ is the contact angle

The effect is known as capillary action or capillarity and is the basic driving force for the standard heat pipe.

For non wetting liquids the curved surface is depressed below the plane liquid surface level, in heat pipes wetting liquids are always used.

2.2.3 Change in vapour pressure at a curved liquid surface.   From Fig. 2.5 it is seen that the vapour pressure at the concave surface is less than at the plane liquid surface by an amount equal to the weight of the vapour column, length h.

This pressure difference is

$$P_c - P_o = g\rho_v h$$

Assuming that ρ_v is constant which is very nearly true.

Now: $\quad (\rho_\ell - \rho_v)\, g\, h \;=\; \dfrac{2\sigma_\ell}{r}$

$\therefore \qquad P_c - P_o \;=\; \dfrac{2\sigma_\ell}{r} \times \dfrac{\rho_v}{\rho_\ell - \rho_v}$

This pressure difference $P_c - P_o$ is small compared to the total capillary head $\dfrac{2\sigma_\ell}{r}$ and may be neglected in heat pipe design.

2.2.4 Measurement of surface tension.

There is a large number of methods for the measurement of surface tension of a liquid, and these are described in the standard texts (2.2) (2.3).   For our present purpose we are interested in $\sigma_\ell \cos \theta$ as a measure of the capillary force.   The simplest measurement is that of capillary rise h in a tube, which gives

$$\sigma_\ell \, \cos \theta \;=\; \frac{\rho_\ell\, g h r}{2}$$

In practical heat pipe design it is also necessary to know r, the effective pore radius.   This is by no means easy to estimate for a wick made up of a sintered porous structure or from several layers of gauze.   By measuring the maximum height the working fluid will attain it is possible to obtain information on the capillary head for fluid wick combinations.   Several workers have reported measurements on maximum height for different structures and some results are given in Chapter 3.   The results may differ for the same structure depending on whether the film was rising or falling; the reason for this is brought out in Fig. 2.6.

Interface with
rising column

Interface with
falling column

Fig. 2.6 Rising and falling column interfaces

Another simple method, for the measurement of σ_ℓ, sometimes employed, is the
measurement of maximum bubble pressure, due to Jäger, Fig. 2.7.   The
pressure is progressively increased until the bubble breaks away and the
pressure falls.   When the bubble radius reaches that of the tube pressure
is a maximum P_{max} and at this point

$$P_{max} \; = \; \rho_\ell hg + \frac{2\sigma_\ell}{r}$$

This method was used (by Bohdanski and Schins) for liquid metals (2.4)

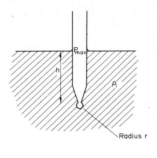

$$\sigma_i = r/2 \; [P_{max} - \rho_l \; gh]$$

Fig. 2.7   Jäger's method for surface tension
         measurement

The surface tension of two liquids can be compared by comparing the mass of
the droplets falling from a narrow vertical tube.   If the masses are m_1,
m_2 respectively

$$\frac{m_1}{m_2}\frac{\rho_{\ell 1}}{\rho_{\ell 2}} \; \simeq \; \frac{\sigma_{\ell 1}}{\sigma_{\ell 2}} \qquad\qquad \ldots \; 2.2.2$$

2.2.5 **Temperature dependence of surface tension.**   Surface tension decreases
as the temperature increases; this effect was first studied by Eötvös (2.3)
and his equation was later modified by Ramsay and Shields to give

$$\sigma_\ell \; (\frac{M}{\rho_\ell})^{\frac{2}{3}} \; = \; H \; (T_c - 6 - T) \qquad\qquad \ldots \; 2.3.3$$

M is the molecular weight

T_c is the critical temperature K

T is the temperature K

H is a constant equal to 2.12 for normal non-associated
   liquids.   Substances such as water and alcohols
   containing hydroxyl groups give a lower value of H,
   which is also temperature dependent

The Eötvös-Ramsay-Shields equation does not give agreement with the experimentally observed behaviour of liquid metal and molten salts.

Bohdanski and Schins (2.5) have derived an equation:

$$\frac{\sigma_\ell}{\sigma_{\ell o}} = 1 - \frac{T}{T_c} - 0.40 \left\{1 - \frac{T}{T_c}\right\} \frac{T}{T_c} \qquad \dots 2.2.4$$

which gives good agreement with expected results if T_c is 10% less than the true critical temperature.

2.2.6 Capillary pressure ΔP_c.   Equation (2.1) shows that the pressure drop across a curved liquid interface is

$$\Delta P = \frac{2\sigma_\ell}{R}$$

From Fig. 2.8 we see that R cos θ = r
where r is the effective radius of the wick pores and θ the contact angle.
Hence the capillary head at the evaporator

$$\Delta P_e' = 2\sigma_\ell \frac{\cos \theta_e}{r_e}$$

Similarly at the condenser

$$\Delta P_c' = 2\sigma_\ell \frac{\cos \theta_c}{r_c}$$

The resultant capillary head will be

$$\Delta P_c \ = \ 2\sigma_\ell \ (\frac{\cos \theta_e}{r_e} \ - \ \frac{\cos \theta_c}{r_c}) \qquad\qquad ... \ 2.2.5$$

This will have a maximum value when $\cos \theta_e = 1$ and $\cos \theta_c = 0$

Hence $\qquad (\Delta P_c)_{max} \ = \ \frac{2\sigma_\ell}{r_e} \qquad\qquad\qquad ... \ 2.2.6$

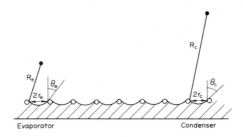

Fig. 2.8   Wick and pore parameters in
evaporator and condenser

2.3 Pressure Difference due to Friction Forces

In this section we will consider the pressure differences in the liquid and
vapour phases which are caused by frictional forces.   It is convenient here
to define some of the terms which will be used later in the chapter.

2.3.1 Laminar flow - the Hagen-Poiseuille equation.

The steady state
laminar flow of an incompressible fluid of constant viscosity μ, through a
tube of circular cross section, radius a, is described by the Hagen-
Poiseuille equation.

This equation relates the velocity v_r of the fluid, at radius r, to the
pressure difference $P_2 - P_1$ across a tube length ℓ.

$$v_r \ = \ \frac{a^2}{4\mu} \left[1 - (\frac{r}{a})^2 \right] \frac{P_2 - P_1}{\ell}$$

The velocity varies in a parabolic way from a maximum value

$$v_m = \frac{a^2}{4\mu} \frac{P_2 - P_1}{\ell}$$

on the axis of the tube to zero at the wall.   The average velocity is given
by

$$v = \frac{a^2}{8\mu} \frac{P_2 - P_1}{\ell}$$

or rearranging

$$\frac{P_2 - P_1}{\ell} = \frac{8\mu v}{a^2} \qquad \ldots 2.3.1$$

In the one dimensional treatments the average velocity v will be used
throughout.

The volume flowing per second S is

$$S = \rho v \pi a^2 = \frac{\pi a^4}{8\mu} \frac{P_2 - P_1}{\ell}$$

and if ρ is the fluid density, the mass flow $\dot{m}$ is given by

$$\dot{m} = \rho S = \frac{\pi a^4}{8\mu} \frac{P_2 - P_1}{\ell} \qquad \ldots 2.3.2$$

2.3.2 Axial Reynold's number, Re.   As the velocity is increased the flow
will change from laminar to turbulent, the transition point is specified in
terms of the Reynold's number Re where

$$Re = \frac{\rho v d}{\mu} \qquad \ldots 2.3.3$$

where d = 2a the pipe diameter.

Re is dimensionless and by writing in the form

$$Re = \frac{\rho v^2}{\mu \frac{v}{d}}$$

is seen to be a measure of the ratio of the inertial to the viscous forces acting on the fluid.

At Re > 2100 the flow pattern changes from laminar to turbulent and the velocity profile alters from the parabolic form to the profile shown in Fig. 2.9.

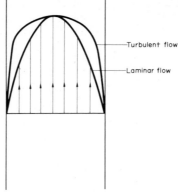

Turbulent flow

Laminar flow

Fig. 2.9 Velocity distribution in a circular tube for laminar and turbulent flow

It is instructive to compare the kinetic head or flow energy to the energy lost by viscous friction over the pipe length ℓ.   We will express both in terms of effective pressure difference, ΔP.

The kinetic energy term $\Delta P_{KE} = \frac{\rho v^2}{2}$

And the viscous term ΔP_F is given by equation 2.2.1

$$\Delta P_F = \frac{8 \mu v \ell}{a^2}$$

$$\frac{\Delta P_{KE}}{\Delta P_{F}} = \frac{\rho v a}{16 \mu \ell}$$

$$= \frac{R_e \cdot a}{32 \ell}$$

That is, assuming the flow still remains laminar, the kinetic energy term ΔP_{KE} is equal to the viscous term ΔP_{F} for a length $\ell = \frac{R_e \cdot a}{32}$ ... 2.3.4

2.3.3 Turbulent flow - the Fanning equation.

The pressure drop for turbulent flow is usually related to the average velocity by the Fanning equation:

$$\frac{P_2 - P_1}{\ell} = \frac{4}{d} \; f \; \frac{1}{2} \; \rho v^2 \qquad \qquad \text{... 2.3.5}$$

where f is the Fanning Factor.

f is related to Reynold's number and in the turbulent region is given by the Blasius equation:

$$f = \frac{0.0791}{R_e^{\frac{1}{4}}} , 2100 < R_e < 10^5 \qquad \qquad \text{... 2.3.6}$$

We see that if we write $f = \frac{16}{R_e}$ for $R_e < 2100$ the Fanning equation reduces to the Hagen-Poiseuille form.

2.3.4 Navier-Stokes equation.

The simple one dimensional treatment given above is usually adequate to describe the situation in the liquid phase. The vapour phase is however more complicated since the radial velocity components in the evaporator and condenser regions should be taken into account.   When this is done it is found that the velocity profile approximates to that for Hagen Poiseuille flow in the evaporator and adiabatic regions but deviates quite considerably from this in the condenser region.   In order to carry out a complete analysis it is necessary to solve the full momentum equation.   This may be expressed in words for a small volume element as

Mass of element × acceleration = Sum of forces acting on the element

= Pressure force + viscous force + gravitational force + other body forces (e.g. magnetic forces)

This equation is known as the Navier-Stokes equation and is written out in its co-ordinate form in Appendix 3.   The Navier-Stokes equation is solved in association with the continuity equation and if the fluid is not incompressible the Equation of State is also required.

The continuity equation may be expressed:

Rate of mass entering the element - Rate of mass leaving the element

= Rate of mass accumulation

The continuity equation is also written out in co-ordinate form in Appendix 3.

The equation of state relates the density of a fluid to its pressure.   We can usually assume that the vapour behaves as a perfect gas, in which case

$$\rho = \frac{P}{RT}$$

where P is the pressure
R the gas constant for the material $= \dfrac{R_o}{M}$
T the absolute temperature

A number of analyses of this type are reported in the literature and their conclusions are discussed in Section 2.5.

2.4 Pressure Difference in the Liquid Phase ΔP_1

The flow regime in the liquid phase is almost always laminar.   Since the liquid channels will not in general be straight nor of circular cross-section and will often be interconnected the Hagen-Poiseuille equation must be modified to take account of these differences.

Since mass flow will vary in both the evaporator and the condenser region an effective length rather than the geometrical length must be used for these regions.   If the mass change per unit length is constant the total mass flow will increase, or decrease, linearly along the regions.   We can therefore replace the lengths of the evaporator ℓ_e and the condenser ℓ_c by $\frac{\ell_e}{2}$ and $\frac{\ell_c}{2}$.   The total effective length for fluid flow will then be ℓ_{eff} where

$$\ell_{eff} = \ell_a + \frac{\ell_e + \ell_c}{2} \qquad \ldots 2.4.1$$

Tortuosity within the capillary structure must be taken into account separately and is discussed below.

There are three principal capillary geometries

(1)   Wick structures consisting of a porous structure made up of interconnecting pores.   Gauzes, felts and sintered wicks come under this heading, these are frequently referred to as homogeneous wicks.

(2)   Open grooves.

(3)   Covered channels consisting of an area for liquid flow closed by a finer mesh capillary structure.   Grooved heat pipes with gauze covering the groove and arterial wicks are included in this category.   These wicks are sometimes described as composite wicks.

2.4.1 Homogeneous wicks.   If ε is the fractional void of the wick, that is the fraction of cross section available for the fluid the total flow cross sectional area is $2\pi (r_w^2 - r_v^2)\varepsilon$

If r_c is the effective pore diameter the Hagen-Poiseuille equation (2.3.2) gives

$$\dot{m} = \frac{2\pi (r_w^2 - r_v^2) \varepsilon r_c^2 \rho_\ell}{8\mu_\ell} \quad \frac{\Delta P_\ell}{\ell_{eff}} \qquad \ldots 2.4.2$$

or expressing it in terms of heat flow $Q = \dot{m}\,L$.
where L is the latent heat or enthalpy of vaporisation
and rearranging

$$\Delta P_\ell \;=\; \frac{8\mu_\ell\,Q\,\ell_{eff}}{2\pi\,(r_w{}^2 - r_v{}^2)\,\epsilon r_c{}^2 \rho_\ell L} \qquad \ldots 2.4.3$$

For porous media this equation is usually written

$$\Delta P_\ell \;=\; \frac{b\mu_\ell\,Q\,\ell_{eff}}{2\pi\,(r_w{}^2 - r_v{}^2)\,\epsilon r_c{}^2 \rho_\ell L} \qquad \ldots 2.4.4$$

where b is a dimensionless constant of about 10 - 20 to include a correction
for tortuosity.

Whilst this relation can be useful for theoretical treatment it contains
three constants b, ϵ and r_c which are all difficult to measure in practice.
It is therefore more usual for wick structure of this type to use Darcy's
'Law' for the calculation of ΔP_ℓ.

Darcy's 'Law' is written

$$\Delta P_\ell \;=\; \frac{\mu_\ell\,\ell_{eff}\,\dot{m}}{\rho_\ell\,KA} \qquad \ldots 2.4.5$$

   where   K is the wick permeability
            A is the wick cross sectional area

Comparing (2.4.5) with equation (2.4.2) we see that Darcy's 'Law' is the
Hagen Poiseuille equation with correction terms included in K to take account
of pore size, distribution and the tortuosity.   It serves as a definition of
permeability K a quantity which is easily measured.

The Blake-Kozeny equation is sometimes used in the literature.   This
equation relates the pressure gradient across a porous body, made up from
spheres diameter D, to the flow of liquid.   Like Darcy's 'Law' it is merely
the Hagen-Poiseuille equation with correction factors.   The Blake-Kozeny

equation may be written

$$\Delta P_\ell = \frac{150 \mu (1 - \epsilon')^2 \ell_{eff} v}{D^2 \epsilon'^3} \qquad \ldots 2.4.6$$

and is applicable only to laminar flow, which requires

$$R_e' = \frac{\rho v D}{\mu(1 - \epsilon')} < 10$$

where v is the superficial velocity $\frac{\dot{m}}{\rho_\ell A}$

and ϵ' is $\dfrac{\text{volume of voids}}{\text{volume of body}}$

A number of wick sections are shown in Fig. 2.10.

2.4.2 Longitudinal groove wick. For grooved wicks the pressure drop in the liquid is given by:

$$\Delta P_\ell = \frac{8\mu_\ell Q\ell}{\pi r_e^4 N \rho_\ell L} \qquad \ldots 2.4.7$$

where N is the number of grooves

r_e is the effective groove radius and r_e is defined by the effective hydraulic radius relationship

$$r_e = 2 \frac{\text{Flow area}}{\text{Wetted perimeter}} \qquad \ldots 2.4.8$$

At high vapour velocities shear forces will tend to impede the liquid flow in open grooves. This may be avoided by using a fine pore screen to form a composite wick structure.

2.4.3 Composite wicks. Such a system as arterial or composite wicks require an auxiliary capillary structure to distribute the liquid over the evaporator and condenser surfaces.

$$\Delta P_\ell = \frac{8\mu_\ell Q\ell}{\pi r^4 \rho_\ell L} \qquad \ldots 2.4.9$$

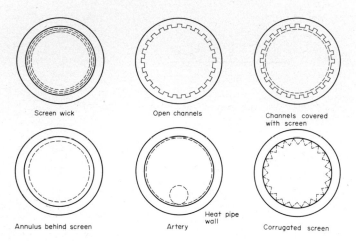

Fig. 2.10 A selection of wick sections

The pressure drop in wicks constructed by an inner porous screen separated from the heat pipe wall to give an annular gap for the liquid flow, may be obtained from the Hagen-Poiseuille equation applied to parallel surfaces, provided that the annular width w is small compared to the radius of the pipe vapour space r_v.

In this case:

$$\Delta P_\ell = \frac{6\mu_\ell Q\ell}{\pi r_v w^3 \rho_\ell L} \qquad \qquad \text{... 2.4.10}$$

This wick structure is particularly suitable for liquid metal heat pipes.

Crescent annuli may be used, in which it is assumed that the screen is moved down to touch the bottom of the heat pipe wall leaving a gap 2w at the top. In this case:

$$\Delta P_\ell = \frac{2.4\mu_\ell Q\ell}{\pi r_v w^3 \rho_\ell L} \qquad \qquad \text{... 2.4.11}$$

In equations 2.4.9 - 2.4.11, the length of the heat pipe should be taken as l_{eff}, defined in eqn. 2.4.1.

2.5 Vapour Phase Pressure Difference, ΔP_v

2.5.1 **Introduction.**   The total vapour phase difference in pressure will be
the sum of the pressure drops in the three regions, namely the evaporator
drop ΔP_{ve}, the adiabatic section drop ΔP_{va}, and the pressure drop in the
condensing region ΔP_{vc}.

The problem of calculating the vapour pressure drop is complicated in the
evaporating and condensing regions by radial flow due to evaporation or
condensation.   It is convenient to define a further Reynolds number, the
Radial Reynolds Number $R_r = \dfrac{\rho_v v r_v}{\mu_v}$ to take account of the radial velocity
component v at the wick $r = r_v$.

By convention the vapour space radius r_v is used rather than the vapour
space diameter which is customary in the definition of axial Reynolds
Number. R_r is positive in the evaporator section and negative in the
condensing section.   In most practical heat pipes R_r lies in the range
0.1 to 100.

R_r is related to the radial rate of mass injection or removal per unit
length $\dfrac{d\dot{m}}{dz}$ as follows $R_r = \dfrac{1}{2\pi\mu_v}\dfrac{d\dot{m}}{dz}$ ... 2.5.1

The radial and axial Reynold's numbers are related for uniform evaporation
or condensation rates, by the equation,

$$R_r = \frac{R_e}{4}\frac{r_v}{z}$$ ... 2.5.2

where z is the distance from either the end of the evaporator section or
the end of the condenser section.

In Section 3 we showed in Equation 2.3.4, that provided the flow is laminar,
the pressure drop due to viscous forces in a length ℓ is equal to the
kinetic head when

$$\ell = \frac{R_e a}{32} = \frac{R_e r_v}{32}$$

If we substitute $R_e = \dfrac{4R_r \ell}{r_v}$ for the evaporator or condenser region we find that the condition reduces to

$$R_r = 8 \qquad\qquad\qquad \ldots 2.5.3$$

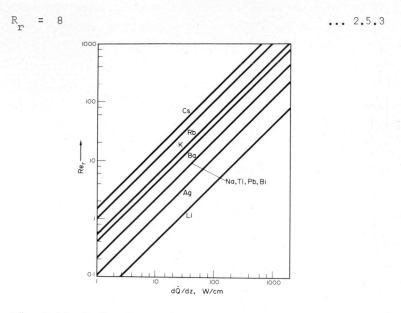

Fig. 2.11 Radial Reynolds number versus heat input per unit length of heat pipe, (liquid metal working fluids)

Fig. 2.11 taken from Busse (2.6) shows R as a function of power/unit length for various liquid metal working fluids.

2.5.2 Incompressible flow: (simple one dimensional theory). In the following treatment we will regard the vapour as an incompressible fluid. This assumption implies that the flow velocity v is small compared to the velocity of sound C, in the vapour i.e. the Mach number $\frac{v}{C} < 0.3$. Or expressed in a different way we are considering heat pipes in which ΔP_v is small compared with P_v the average vapour pressure in the pipe. This assumption is not valid during startup nor is it always true in the case of high temperature liquid metal heat pipes. The effect of compressibility of the vapour will be considered in a later section.

In the evaporator region the vapour pressure gradient will be necessary to carry out two functions.

(i)   To accelerate the vapour entering the evaporator section up to the
      axial velocity v.   Since on entering the evaporator this vapour
      will have radial velocity but no axial velocity.   The necessary
      pressure gradient we will call the inertial term $\Delta P_v{}'$

(ii) To overcome frictional drag forces at the surface $r = r_v$ at the
      wick.   This is the viscous term $\Delta P_v{}''$

We can estimate the magnitude of the inertial term as follows.   If the mass
flow/unit area of cross-section at the evaporator is ρv then the corresponding
momentum flux/unit area will be given by $\rho v \times v$ or ρv^2.   This momentum flux
in the axial direction must be provided by the inertial term of the pressure
gradient.

Hence
$$\Delta P_v{}' = \rho_v{}^2 \qquad \ldots 2.5.4$$

Note that $\Delta P_v'$ is independent of the length of the evaporator section.   The
way in which $\Delta P_v{}'$ varies along the length of the evaporator is shown in
Fig. 2.12(a).

If we assume laminar flow we can estimate the viscous contribution to the
total evaporator pressure loss by integrating the Hagen-Poiseuille equation.
If the rate of mass entering the evaporator per unit length dm/dz is constant
we find by integrating Equation 2.3.2 along the length of the evaporator
section

$$\Delta P_v{}'' = \frac{8\mu_v \dot{m}}{\pi r_v{}^4} \frac{\ell_e}{2} \qquad \ldots 2.5.5$$

Thus the total pressure drop in the evaporator region ΔP_{ve} will be given by
the sum of these two terms

$$\Delta P_{ve} = \Delta P_v{}' + \Delta P_v{}''$$

$$= \rho_v{}^2 + \frac{8\mu_v \dot{m}}{\pi r_v{}^4} \frac{\ell_e}{2} \qquad \ldots 2.5.6$$

The condenser region may be treated in a similar manner, but in this case
axial momentum will be lost as the vapour stream is brought to rest so the
inertial term will be negative, that is there will be pressure recovery.
For the simple theory the two inertial terms will cancel and the total
pressure drop in the vapour phase will be due entirely to the viscous terms.
It is shown later that it is not always possible to recover the inertial
pressure term in the condensing region.

In the adiabatic section the pressure difference will contain only the
viscous term which will be given either by the Hagen-Poiseuille equation or
the Fanning equation depending on whether the flow is laminar or turbulent.

For laminar flow

$$\Delta P_a = \frac{8\mu_v \dot{m}}{\pi r_v^4} \ell_a \qquad\qquad R_e < 2100 \qquad\qquad \ldots 2.5.7$$

For turbulent flow

$$\Delta P_a = \frac{2}{r_v} f \frac{1}{2} \rho_v \times v^2 \ell_a \qquad\qquad R_e > 2100 \qquad\qquad \ldots 2.5.8$$

where $\quad f = \dfrac{0.0791}{R_e^{\frac{1}{4}}}$

Hence the total vapour pressure drop ΔP_v is given by:

$$\Delta P_v = \Delta P_{ve} + \Delta P_{vc} + \Delta P_{va}$$

$$= \rho_v^2 + \frac{8\mu_v \dot{m}}{\pi r_v^4} \left[\frac{\ell_e + \ell_c}{2} + \ell_a \right] \qquad\qquad \ldots 2.5.9$$

for laminar flow and no pressure recovery

$$\Delta P_v = \frac{8\mu_v \dot{m}}{\pi r_v^4} \left[\frac{\ell_e + \ell_c}{2} + \ell_a \right] \qquad\qquad \ldots 2.5.10$$

for laminar flow with full pressure recovery.

Equations 2.5.9 and 2.5.10 enable the calculation of vapour pressure drop in simple heat pipe design and are used extensively.

2.5.3 Incompressible flow - one dimensional theories of Cotter and Busse.

In addition to the assumption of incompressibility the above treatment assumes a fully developed flow velocity profile and complete pressure recovery.   It is however broadly correct.   A considerable number of papers have been published giving a more precise treatment of the problem.   Some of these will now be summarised in this and the following section.

The earliest theoretical treatment of the heat pipe was by Cotter.   For R_r << 1 Cotter used the following result obtained by Yuan and Finkelstein for laminar incompressible flow in a cylindrical duct with uniform injection or suction through a porous wall (2.1).

$$\frac{dP_v}{dz} = - \frac{8\mu_v \dot{m}}{\pi\rho_v r_v^4}\left[1 + \tfrac{3}{4} R_r - \tfrac{11}{270} R_r^2 \cdots\cdots\right]$$

He obtained the following expression

$$\Delta P_{ve} = - \frac{4\mu_v \ell_e Q}{\pi\rho_v r_v^4 L} \qquad\qquad \cdots 2.5.11$$

which is identical to Equation 2.5.5

For R_{re} >> 1 Cotter used the pressure gradient obtained by Knight and McInteer for flow with injection or suction through porous parallel plane walls.   The resulting expression for pressure gradient is

$$\Delta P_{ve} = - \frac{\dot{m}^2}{8\rho_v r_v^4}$$

Rewriting this in the form

$$\Delta P_{ve} = - \frac{(\rho_v \pi r_v^2 v)^2}{8\rho_v r_v^4} = - \frac{\pi^2}{8}\rho_v v^2 \qquad\qquad \cdots 2.5.12$$

$$\Delta P_{ve} = -\rho_v v^2 \qquad\qquad \ldots 2.5.4$$

it is seen not to differ greatly from the Equation 2.5.4 obtained previously.

However in the condenser region Cotter used a different velocity profile and obtained a pressure recovery value of

$$\Delta P_{vc} = + \frac{4}{\pi^2} \frac{\dot{m}^2}{8\rho_v r_v^4} \qquad\qquad \ldots 2.5.13$$

which is only $\frac{4}{\pi^2}$ of ΔP_{ve}, giving only partial pressure recovery.

In the adiabatic region Cotter assumed fully developed flow and hence a value for

$$\Delta P_{va} = - \frac{8\mu_v \dot{m} \ell_a}{\pi \rho_v r_v^4} \qquad\qquad \ldots 2.5.14$$

Cotter's full expression for

$$\Delta P_v = - (1 - \frac{4}{\pi^2}) \frac{\dot{m}^2}{8\mu_v r_v^4} - \frac{8\mu_v \dot{m} \ell_a}{\pi \rho_v r_v^4} \qquad\qquad \ldots 2.5.15$$

Busse also considered the one dimensional case, assuming a modified Hagen-Poiseuille velocity profile and using this to obtain a solution of the Navier-Stokes equation for a long heat pipe.   Busse's results are as follows

$$\Delta P_{ve} = - \frac{8\mu_v \dot{m}}{\pi \rho_v r_v^4} \frac{\ell_e}{2} \left[1 + R_{re} \frac{7}{9} - \frac{8A}{27} + \frac{23A^2}{405} \right] \ldots 2.5.16$$

$$\Delta P_{vc} = - \frac{8\mu_v \dot{m}}{\pi \rho_v r_v^4} \frac{\ell_c}{2} \left[1 - R_{rc} \frac{7}{9} - \frac{8a}{27} + \frac{23a^2}{405} \right] \ldots 2.5.17$$

$$\Delta P_{va} = - \frac{8\mu_v \dot{m} \ell_a}{\pi \rho_v r_v^4} \left[1 + \frac{R_e r_v}{8\ell_a} \frac{8}{27} (A-a) - \frac{23}{405} (A-a)^2 \right]$$

$$\ldots 2.5.18$$

We recognise the first term in each of the above equations as the viscous contribution to pressure gradient arising in laminar flow.   The remaining terms take account of inertial effects.   Consider these terms for ΔP_{ve}.

i.e. $- \dfrac{8\mu_v \dot{m}_v}{\pi\rho_v r_v^4} \dfrac{\ell_e}{2} \dfrac{1}{2\pi\mu_v} \dfrac{\dot{m}_v}{\ell_e} \left[\dfrac{7}{9} - \dfrac{8A}{27} + \dfrac{23A^2}{405} \right]$

The factor $\dfrac{7}{9} - \dfrac{8A}{27} - \dfrac{23A^2}{405}$ ranges from 0.78 for $R_r = 0$ to 0.61 for $R_r = \infty$ and the expression is seen to be of the same form as Equation 2.5.4 and not very different from it.

The total vapour pressure difference is

$$\Delta P_v = \Delta P_{ve} + \Delta P_{vc} + \Delta P_{va}$$

$$= - \dfrac{8\mu_v \dot{m}_v}{\rho_v r_v^4} \left[\ell_a + \dfrac{\ell_e + \ell_c}{2} \right] \qquad \text{... 2.5.19}$$

and the inertial terms cancel out implying full pressure recovery.

2.5.4 Pressure recovery.

We have seen that the pressure drop in the evaporator and condenser regions consists of two terms, an inertial term and a term due to viscous forces.   Simple theory suggests that the inertial term will have the opposite sign in the condenser region and should cancel out that of the evaporator, Fig. 2.12(b).   There is experimental evidence for this pressure recovery for example the photograph by Kemme of a sodium heat pipe provides an impressive demonstration.   In these experiments Kemme achieved 60% pressure recovery.   The radial Reynold's number was greater than 10, Fig. 2.13.   For simplicity in Fig. 2.12(b) we have omitted the viscous pressure drop.   The liquid pressure drop is also shown in Fig. 2.14(a).   Ernst has pointed out that if the pressure recovery in the condenser region is greater than the liquid pressure drop, Fig. 2.14(b) then the meniscus in the wick will be convex.   Whilst this is possible in principle under normal heat pipe operation there is excess liquid in the condenser region so that this condition cannot occur.   For this reason if $\Delta P_{vc} > \Delta P_{\ell c}$ it is usual to neglect pressure recovery and assume that there is no resultant pressure drop or gain in the condenser region, Fig. 2.14(a).

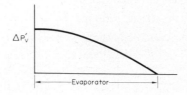

(a)   Pressure variation along the evaporator

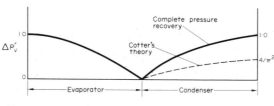

(b)   Pressure recovery of inertial pressure term
        (Adiabatic section omitted)

Fig. 2.12   Vapour pressure change due to
inertial effects

Fig. 2.13   Sodium heat pipe constructed by Kemme
(2.7)

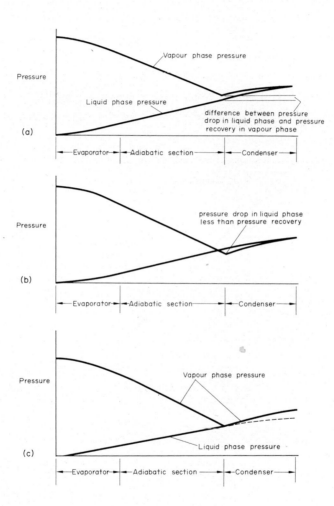

Fig. 2.14   Pressure recovery in a heat pipe (2.8)

2.5.5 Two dimensional incompressible flow.   The previous discussion has
been restricted to one dimensional flow.   In practical heat pipes the
temperature and pressure are not constant across the cross section and this
variation is particularly important in the condenser region.   A number of
authors have considered this two dimensional problem.   Bankston and Smith
and also Rohani and Tien have solved the Navier Stokes equation by numerical
methods.   Bankston and Smith showed that axial velocity reversal assured
at the end of the condenser section under conditions of high evaporation and
condensation rates.   Reverse flow occurs for R_r < -2.3.   In spite of this
reverse flow Busse's simple one dimensional treatment agrees well with the
results of Bankston and Smith up to R_r > 10 in the condenser region, (2.9)
Fig. 2.15.

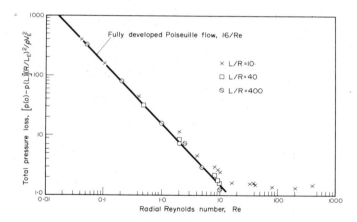

Fig. 2.15 Pressure loss for symmetrical heat pipes
compared with that if the flow were Poiseuille

Rohani and Tien consider the case of a sodium heat pipe of dimensions

$$
\begin{aligned}
Length &= 6 \text{ m} \\
\ell_e &= 0.2 \text{ m} \\
\ell_c &= 0.3 \text{ m} \\
r_v &= 0.0086 \text{ m}
\end{aligned}
$$

The following cases were examined (2.10).

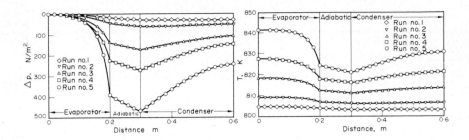

Fig. 2.16 Axial pressure
drop along a heat pipe

Fig. 2.17 Axial temperature
drop along a heat pipe

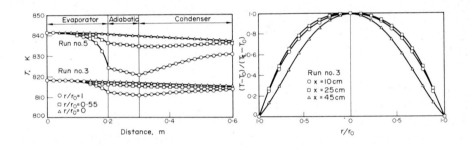

Fig. 2.18 Axial temperature
variation at various radial
locations

Fig. 2.19 Radial temperature
variation at various axial
locations

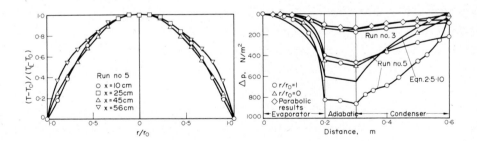

Fig. 2.20 Radial
temperature variation at
various axial locations

Fig. 2.21 Comparison of
elliptic, parabolic and
simple theory pressure
drop results

TABLE 2.1 PRIMARY INFORMATION ON THE SODIUM LIQUID
METAL HEAT PIPE

Run No.	T_{ac} (K)	T_{ae} (K)	H_c (Wm^{-2}K^{-1})	H_e (Wm^{-2}K^{-1})	Re_e	Re_c	T_o(K)	p_o(N/m^2)
1	800	805	17106	2851	2	1.33	804	946
2	800	810	17106	2851	4	2.66	808	1023
3	800	820	17106	2851	8	5.33	816	1203
4	800	830	17106	2851	24	16.00	824	1398
5	800	845	17106	2851	36	24.00	836	1633

TABLE 2.2 COMPARISON OF THE ACTUAL AND EXPECTED
VALUES OF Q AND Re_{max}

Run No.	Q_n(W)	$Re_{max, n}$	Q(W)	Re_{max}
1	185	186	162	163
2	370	372	289	292
3	740	744	610	614
4	1120	1126	906	911
5	1665	1674	1265	1283

2.5.6 Compressible flow.

So far we have neglected the effect of compressibility of the vapour on the operation of the heat pipe. Compressibility can be important during start-up and also in high temperature liquid metal heat pipes; it is discussed in this section.

In a cylindrical heat pipe the axial mass flow increases along the length of the evaporator region to a maximum value at the end of the evaporator; it will then decrease along the condenser region. The flow velocity will rise to a maximum value at the end of the evaporator region where the pressure will have fallen to a minimum.   Deverall et al have drawn attention to the similarity in flow behaviour between such a heat pipe and that of a gas flowing through a converging diverging nozzle.   In the former the area remains constant but the mass flow varies, in the latter the mass flow is

constant but the cross-sectional area is changed.   It is helpful to examine
the behaviour of the convergent-divergent nozzle in more detail before
returning to the heat pipe.   Let the pressure of the gas at the entry to
the nozzle be kept constant and consider the effect of reducing the pressure
at the outlet.   Referring to Fig. 2.22 Curve A.   Here the pressure
difference between inlet and outlet is small.   The gas velocity will
increase to a maximum value in the position of minimum cross-section, or
throat, rising again in the divergent region.   The velocity will not reach
the sonic value.   The pressure passes through a minimum in the throat.   If
the outlet pressure is now reduced the situation shown in Curve B can be
attained.   Here the velocity will increase through the convergent region,
rising to the sonic velocity in the throat.   As before, the velocity will
reduce during travel through the diverging section and there will be some
pressure recovery (2.11).

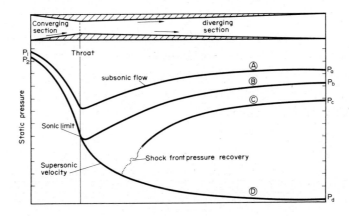

Fig. 2.22 Pressure profiles in a converging –
diverging nozzle

If the outlet pressure is further reduced Curve C the gas will continue to
accelerate after entering the divergent region and will become supersonic.
Pressure recovery will occur after a shock front.

Curve D shows that for a certain exit pressure the gas can be caused to
accelerate throughout the diverging region.   Further pressure reduction will
not affect the flow pattern in the nozzle region.   It should be noted that

after Curve C pressure reduction does not affect the flow pattern in the
converging section, hence the mass flow does not increase after the throat
velocity has attained the sonic value.   This condition is referred to as
choked flow.

Kemme has shown very clearly that a heat pipe can operate in a very similar
manner to the diverging converging nozzle.   His experimental arrangement is
shown in Fig. 2.23.   Kemme used sodium as the working fluid and maintained
a constant heat input of 6.4 kW.   He measured the axial temperature
variation, but since this is related directly to pressure his temperature
profile can be considered to be the same as the pressure profile.   Kemme
arranged to vary the heat rejection at the condenser by means of a gas gap
whose thermal resistance could be altered by varying the Argon-Helium ratio
of the gas.   Kemme's results are shown in Fig. 2.23.   Curve A demonstrates
subsonic flow with pressure recovery; Curve B, obtained by lowering the
condenser temperature, achieved sonic velocity at the end of the evaporator,
and hence operated under choked flow conditions.   Further decrease in the
thermal resistance between the condenser and heat sink simply reduced the
condenser region temperature but did not increase the heat flow which was
limited by the choked flow condition and fixed axial temperature drop in the
evaporator (2.13).

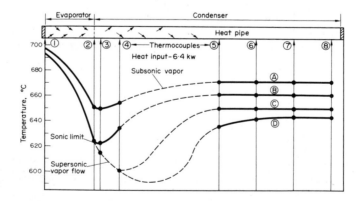

Fig. 2.23   Temperature profiles in a heat pipe (2.12)

It should be noted that under these conditions of sonic limitation considerable axial temperature and pressure changes will exist and the heat pipe operation will be far from isothermal.

Deverall et al have shown that a simple one dimensional model provides a good description of the compressible flow behaviour. Consider the evaporator section (2.11).

The pressure drop $P_o - P_1$ along the evaporator will be given by Equation 2.5.4.

$$P_o - P_1 = \rho v^2 \qquad \ldots 2.5.4$$

Now the equation of state may be written

$$P = \rho_v RT \qquad \ldots 2.5.20$$

also $\quad C = \sqrt{\gamma RT}$

and $\quad M = \dfrac{v}{C} \qquad \ldots 2.5.21$

Substituting into Equation 2.5.4

$$\frac{P_o}{P_1} - 1 = \frac{\rho v^2}{P_1} = \frac{M^2 \gamma RT_1}{RT_1} = \gamma M^2 \qquad \ldots 2.5.22$$

T_o may be regarded as the total head temperature.

Hence
$$\dot{m}\, C_p\, T_o = \dot{m}\, (C_p\, T_1 + \frac{v^2}{2})$$

or $\quad \dfrac{T_o}{T_1} = 1 + M^2\, \dfrac{\gamma - 1}{2} \qquad \ldots 2.5.23$

The density ratio $\quad \dfrac{\rho_o}{\rho_1} = \dfrac{P_o}{P_1}\dfrac{T_1}{T_o} = \dfrac{1 + \gamma\, M^2}{1 + \frac{\gamma - 1}{2} M^2} \qquad \ldots 2.5.24$

Finally the energy balance in the evaporator section gives

$$Q = \rho_v \, A \, v \, L$$

$$= \rho_v \, A \, M \, C \, L \qquad\qquad ... \; 2.5.25$$

Note C is the velocity of sound corresponding to T_1.   Q is sometimes expressed in terms of the velocity of sound C_o corresponding to T_o.   In this case

$$Q = \frac{\rho_v \, A \, m \, C_o \, L}{\sqrt{2} \, (\gamma + 1)} \qquad\qquad ... \; 2.5.26$$

an expression first obtained by Levy.

Expressions for pressure ratio, temperature ratio and density ratio under choked conditions are obtained by substituting M = 1 into Equations 2.5.22, 2.5.23 and 2.5.24.

$$\frac{P_o}{P_s} = 1 + \gamma = 2.66 \qquad\qquad ... \; 2.5.27$$

$$\frac{T_o}{T_s} = \frac{1 + \gamma}{2} = 1.33 \qquad\qquad ... \; 2.5.28$$

$$\frac{\rho_o}{\rho_s} = 2 \qquad\qquad ... \; 2.5.29$$

where $\gamma = 1.66$ for monatomic vapours.

2.6 Gravitational head

The pressure difference due to the hydrostatic head of the liquid may either be positive, negative or zero, depending on the relative positions of the evaporator and condenser.   This pressure difference ΔP_g is given by the expression

$$\Delta P_g = \rho_\ell \, g \, \ell \, \sin \phi \qquad\qquad ... \; 2.6.1$$

where ρ_ℓ is the liquid density

g is the acceleration due to gravity

ℓ is the heat pipe length

ϕ the angle made by the heat pipe with the horizontal.
(ϕ is positive when the condenser is lower than the
evaporator)

2.7 Entrainment

In a heat pipe the vapour flows from the evaporator to the condenser and the
liquid is returned by the wick structure.   At the interface between the wick
surface and the vapour the latter will exert a shear force on the liquid in
the wick.   The magnitude of the shear force will depend on the vapour
properties and velocity and its action will be to entrain droplets of
liquid and transport them to the condenser end.   This tendency to entrain
is resisted by the surface tension in the liquid.   Entrainment will prevent
the heat pipe operating and represents one limit to performance.   Kemme
observed entrainment in a sodium heat pipe and reports that the noise of
droplets impinging on the condenser end could be heard.

The Weber Number, W_e* which is defined as the ratio between inertial vapour
forces and liquid surface tension forces provides a convenient measure of
the likelihood of entrainment.

$$W_e = \frac{\rho_v v^2 Z}{2\pi\sigma_\ell} \qquad\qquad ... \; 2.7.1$$

where ρ_v is the vapour density

v the vapour velocity

σ_ℓ surface tension

Z a dimension characterising the vapour liquid surface.
In a wicked heat pipe it is related to the wick spacing.

* Some authors define $W_e = \dfrac{\rho_v v^2 Z}{\sigma_\ell}$

If it is assumed that entrainment will occur when W_e = 1, (this assumption
in effect determines Z), the limiting vapour velocity V_c will be given by

$$V_c = \sqrt{\frac{2\pi\sigma_\ell}{\rho_v Z}} \qquad\qquad \ldots 2.7.2$$

Since the axial energy flux

$$q = \rho_v Lv$$

The entrainment limited axial flux is given by

$$q = \sqrt{\frac{2\pi\rho_v L^2 \sigma_\ell}{Z}} \qquad\qquad \ldots 2.7.3$$

$\rho_v L^2 \sigma_\ell$ is a figure of merit for working fluids from the point of view of
entrainment.

Cheung has plotted this figure of merit against temperature for a number of
liquid metals, Fig. 2.24 (2.14).

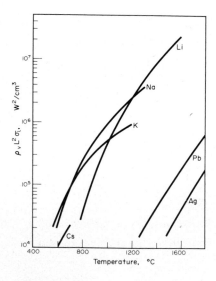

Fig. 2.24   Entrainment merit number (2.14)

A number of authors report experimental results on the entrainment limit.
They usually select a value of Z which fits their results and show that the
temperature dependence of the limit is as predicted by Equation 2.7.3.
Some of these results are discussed in Section 2.9.

2.8   Heat Transfer and Temperature Differences in Heat Pipes

2.8.1   Introduction.

In this section we consider the transfer of heat
and the associated temperature drops in a heat pipe.   The latter can be
represented by thermal resistances and an equivalent circuit is shown in
Fig. 2.25.   Heat can both enter and leave the heat pipe by conduction from
a heat source/sink, by convection, or by thermal radiation.   The pipe may
also be heated by eddy currents or by electron bombardment, and cooled by
electron emission.   Further temperature drops will occur by thermal
conduction through the heat pipe walls at both the evaporator and condenser
regions.   The temperature drops through the wicks arise in several ways
and are discussed in detail later in this section.   It is found that a
thermal resistance exists at the two vapour-liquid surfaces and also in the
vapour column.

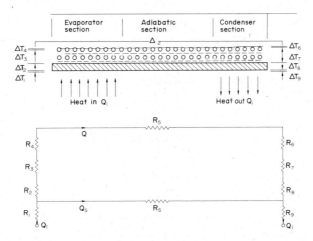

Fig. 2.25   Temperature drops and
equivalent thermal resistances in
a heat pipe

The processes of evaporation and condensation are examined in some detail
both in order to determine the effective thermal resistances and also to
identify the maximum heat transfer limits in the evaporator and condenser
regions.

Finally the results for thermal resistance and heat transfer limits are
summarised for the designer.

2.8.2   **Heat transfer in the evaporator region.**   For low values of heat
flux the heat will be transported to the liquid surface partly by
conduction through the wick and liquid and partly by natural convection.
Evaporation will be from the liquid surface.   As the heat flux is
increased the liquid in contact with the wall will become progressively
superheated and bubbles will form at nucleation sites.   These bubbles
will transport some energy to the surface by latent heat of vaporisation
and will also greatly increase convective heat transfer.   With further
increase of flux a critical value will be reached, burnout, at which the
wick will dry out and the heat pipe will cease to operate.

Before discussing the case of wicked surfaces the data on heat transfer
from plane, unwicked surfaces will be summarised.   Experiments on wicked
surfaces are then described and correlations given to enable the temperature
drop through the wick and the burnout flux to be estimated.   The subject
is complex and further work is necessary to provide an understanding of the
processes in detail.

2.8.3   **Heat transfer from plane surfaces (2.15) (2.16).**   Consider a plane
heater immersed in a pool of liquid, which is maintained at a temperature
T_s, the boiling point corresponding to the pressure of the system.   Let
T_w be the temperature of the heater surface.   Fig. 2.26 shows how the heat
flux q will vary with $T_w - T_s$ as the heater power is raised.   This curve
was first obtained by Nukiyama (2.17) for water and it is found that all
liquids behave in a similar way.

The region A-B corresponds to natural convection to the evaporating surface.

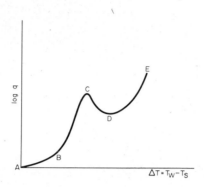

Fig. 2.26   Pool boiling regimes

As the flux is increased bubbles form at the surface and very high heat
transfer rates can be achieved with quite small temperature differences.
This region B-C is known as the nucleate boiling or pool boiling region.
The bubbles both transport heat by latent heat and also increase
convective heat transport.

The heat flux in nucleate boiling cannot be increased indefinitely, at
point C the bubble population becomes so high that it becomes difficult
for the liquid to reach the heater surface and a continuous vapour film
forms.   The temperature difference increases rapidly and the condition
is known as burnout, boiling crisis, or critical heat flux.   The region
C-D is known as the partial film boiling region, boiling is unstable and
the surface is alternately covered by vapour and liquid.   From D-E the
vapour film is stable and the condition called stable film boiling.   The
point E is determined by the melting point of the heater material.

Nucleate boiling and bubble formation (2.18).   Equation 2.2.1 showed that
the pressure difference across a curved surface radius R was given by

$$\Delta P = \frac{2\sigma_\ell}{R} \qquad \qquad \ldots 2.2.1$$

For a bubble to form it must start at a nucleation centre which provides
a finite initial radius.   In addition the liquid must be superheated in
order to provide the pressure difference ΔP.

The amount of superheat required ΔT_s may be related to ΔP by the Clausius-Clapeyron equation:

$$\frac{dP}{dT} = \frac{L}{T(V_v - V_\ell)} \qquad \ldots 2.8.1$$

where V_v is the volume of unit mass of vapour

V_ℓ is the volume of unit mass of the liquid

Normally $V_\ell \ll V_v$

Hence $\frac{dP}{dT} = \frac{L}{TV_v}$

Combining 2.2.1 and 2.8.1

$$\Delta T = \frac{2\sigma_\ell T V_v}{Lr} = \frac{2\sigma_\ell T}{\rho_v Lr} \qquad \ldots 2.8.2$$

The mechanism of bubble formation depends strongly on the wetting characteristics of the heating surface.   Wetting has already been referred to in Section 2.2 and Fig. 2.4.   The effect of wetting on bubble formation is shown in Fig. 2.27a.   Bubbles form most readily if the surface is non-wetting.   In addition to wetting nucleation sites are necessary for bubble formation.

Nucleation sites are provided by scratched or rough surfaces and by the release of absorbed gas.   Fig. 2.27b shows how a bubble forms in a crevasse in a surface.   As might be expected a much higher superheat is required to form bubbles on a clean, smooth surface.   Fig. 2.28 shows how the temperature varies with distance from the surface under nucleate boiling conditions, for the case of water on stainless steel at atmospheric pressure. Schins (2.19) reports similar measurements for sodium boiling at 785°C-875°C in a stainless steel boiler.   Schins found that due to the wetting of stainless steel by the sodium nucleation was difficult to start and a super-heat of 100°C was required.   After some time the surface became conditioned and the superheat dropped to around 20°C, this result is surprising since one might have expected the reverse behaviour.

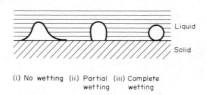

(i) No wetting (ii) Partial (iii) Complete
                    wetting   wetting

(a) Effect or wetting

(b) Bubble formation at a crevasse nucleation site

Fig. 2.27 Bubble formation on a heated surface

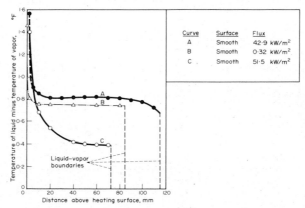

Curve	Surface	Flux
A	Smooth	42·9 kW/m²
B	Smooth	0·32 kW/m²
C	Smooth	51·5 kW/m²

Fig. 2.28 Variation of temperature with distance from
a surface under nucleate boiling considerations (2.15)

Hsu (2.20) has obtained an expression, similar in form to Equation 2.8.2

$$\Delta T = \frac{12.8\sigma_\ell T_S}{J\rho_v L\delta}$$

$$= \frac{3.06\sigma_\ell T_S}{\rho_v L\delta} \qquad \text{in S.I. units} \qquad \ldots 2.8.3$$

where δ is the thermal layer thickness, as a first approximation this may be

60   .P.D. DUNN   D.A. REAY

taken as the average diameter of cavities on the surface.   For typical
surfaces this is around 10^{-3} inches or 2.5×10^{-3} cm.

Table 2.3 lists values of ΔT corresponding to the boiling point at
atmospheric pressure for a number of fluids, (δ = 2.5×10^{-3} cm)

TABLE 2.3   SUPERHEAT ΔT CALCULATED FOR SOME LIQUIDS,
AT THEIR BOILING POINTS AT ATMOSPHERIC PRESSURE, USING
HSU's FORMULA (2.20)

Fluid	Boiling point K	Vapour density kg/m^3	Latent heat kJ/kg	Surface tension N/m	ΔT $^{\circ}$C
Ammonia (NH$_3$)	239.7	0.3	1350	0.028	2.0
Ethyl alcohol (C$_2$H$_5$OH)	338	2.0	840	0.021	0.51
Water	373	0.60	2258	0.059	1.9
Potassium	1047	0.486	1938	0.067	8.9
Sodium	1156	0.306	3913	0.113	26.4
Lithium	1613	0.057	19700	0.26	44.8

Correlation of nucleate boiling data.   Nucleate boiling is very dependent
on the heated surface (2.21) and factors such as release of absorbed gas,
surface roughness, surface oxidation and wettability greatly effect the
surface to bulk liquid temperature difference.   The nature of the surface
may change over a period of time - a process known as conditioning.   (The
effect of pressure is also important.)   For these reasons reproducibility
of results is often difficult.   However a number of authors have proposed
correlations, some empirical and some based on physical models.

Nucleate boiling in water and organic liquids.   A useful correlation
applicable to these fluids has been attained by Rohsenhow (2.22).   In his
model Rohsenhow considered the convective heat transfer due to bubbles.

From dimensional considerations:

$$Nu_b = f(Re_b, Pr_\ell)$$

where: Nu_b is the Nusselt Number for bubbles and is the ratio of heat transferred by convection q to that transmitted by conduction $K_\ell \Delta T$.

i.e. $Nu_b = \dfrac{q}{K_\ell \Delta T}$

Pr_ℓ is the Prandtl Number for the liquid and is defined as $\dfrac{C_p \mu_\ell}{K_\ell}$, the ratio of the kinematic to the thermal diffusivity

Re_b is the Reynolds number for the bubble

$$= \dfrac{\rho_v V_\ell D_\ell}{\mu_\ell} \qquad \text{where } D_\ell \text{ is the bubble diameter}$$

Rohsenhow extended this relation empirically and found:-

$$\frac{Re_b Pr_\ell}{Nu_b} = C\, Re_b{}^n\, Pr_e{}^m$$

Using a 0.024 in. platinum wire in distilled water, he obtained the relation

$$\frac{C_{p\ell}\Delta T}{L} = 0.013 \left[\frac{q}{\mu_\ell h} \sqrt{\frac{g_c}{g}\frac{\sigma}{\rho_\ell-\rho_v}} \right]^{0.33} \left[\frac{C_{p\ell}\mu_\ell}{K_\ell} \right]^{1.7} \qquad \dots 2.8.4$$

The term $\dfrac{g}{g_c}$ is the gravitational acceleration referred to the standard value $g_c = 9.81$ m/s^2 and is required for example in space applications.

The Rohsenhow correlation agrees well with data from other fluid-surface combinations if the constant 0.013 is replaced by values listed in Table 2.4.

TABLE 2.4

Surface fluid combination	C
Water-nickel	0.006
Water-platinum	0.013
Water-copper	0.013
Water-brass	0.006
Water-nickel and stainless steel	0.013
Water-stainless steel	0.014
Carbon tetrachloride-copper	0.013
Benzene-chromium	0.010
n-pentane-chromium	0.015
Ethyl alcohol-chromium	0.0027
Isopropyl alcohol-copper	0.0025
n-butyl alcohol-copper	0.0030

A number of authors have provided relationships to enable the critical heat flux q_c to be predicted.

One of these was developed by Rohsenhow and Griffith (2.23) who obtained the following expression

$$\frac{q_c}{L \cdot \rho_v} = 0.012 \left[\frac{\rho_\ell - \rho_v}{\rho_v} \right]^{0.6} \qquad \ldots \ 2.8.5$$

Another correlation due to Caswell and Balzhieser (2.24) applies to both metals and non metals.

$$\frac{q \ C_p}{L^2 \rho_v K_\ell} \ Pr^{-0.71} = 1.02 \times 10^{-6} \left[\frac{\rho_\ell - \rho_v}{\rho_v} \right]^{0.65} \qquad \ldots \ 2.8.6$$

Other comprehensive references on liquid metal boiling are Subbotin (2.25) and Dwyer, (2.26)

Burn out correlations.   As for nucleate boiling the critical peak flux on boiling crisis flux is also very dependent on surface conditions.   For water at atmospheric pressure the peak flux lies in the range 95 W/cm^2 to 130 W/cm^2 and is between 3 and 8 times the value obtained for organic liquids.   The corresponding temperature difference for both water and organics is between $20^{\circ}C$ and $50^{\circ}C$.

Liquid metals have the advantage of low viscosity and high thermal conductivity and the alkali metals in the pressure range 0.1 - 10 bar give peak flux values of 100 - 300 W/cm^2 and a corresponding temperature difference of around $5^{\circ}C$.

2.8.4 Boiling from wicked surfaces.

There is now a considerable literature on boiling from wicked surfaces.   The work reported includes measurements on plane surfaces and on tubes, the heated surfaces can be horizontal or vertical, and either totally immersed in the liquid or evaporating in the heat pipe mode.   Water, organics and liquid metals have all been investigated.   The effect of the wick is to further complicate the boiling process since in addition to the factors referred to in the section on boiling from smooth surfaces the wick provides sites for additional nucleation and significantly modifies the movement of the liquid and vapour phases to and from the heated surface.

At low values of heat flux the heat transfer is primarily by conduction through the flooded wick.   This is demonstrated in the results of Philips and Hinderman (2.27) who carried out experiments using a wick of 220.5 nickel foam 0.14 cm thick and one layer of stainless steel screen attached to a horizontal plate.   Distilled water was used as the working fluid.   Their results are shown in Fig. 2.29.   The solid curve is the theoretical curve for conduction through a layer of water of the wick thickness.

At higher values of heat flux nucleation occurs.   Ferrell and Alleavitch (2.28) studied the heat transfer from a horizontal surface covered with beds of Monel beads.   Results are reported on 30-40 mesh and 40-50 mesh using water at atmospheric pressure as the working fluid and a total immersion to a depth of 7.5 cm.   The bed depth ranged from 3mm to 25 mm.   They concluded that the heat transfer mechanism was conduction through a saturated wick liquid-matrix to a vapour interface located in the first layer of beads. Agreement was good between the theoretical predictions of this model and the

experimental results.   No boiling was observed.   Fig. 2.30 shows these
results together with the experimental values obtained for the smooth
horizontal surface.   It is seen that the latter agrees well with the
Rohsenhow correlation but that for low values of temperature difference the
heat flux for the wicked surface is much greater than for the smooth
surface.   This effect has been observed by other workers for example
Corman and Welmet (2.29), the curves cross over at higher values of ΔT,
probably because of increased difficulty experienced by the vapour in
leaving the surface.

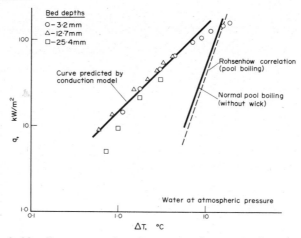

Fig. 2.29   Heat transfer from an submerged wick (2.27)

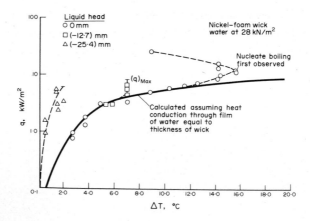

Fig. 2.30   Boiling from submerged wick compared with
boiling from a smooth surface (2.28)

Abhat and Seban (2.30) report a heat transfer measurements on vertical tubes using water, ethanal and acetone as the working fluids.   In this series of experiments results are given for smooth surfaces, immersed wicks, and evaporating wicks.   The results for water are shown in Fig. 2.31.   The authors conclude that up to heat fluxes of 15 W/cm^2 the heat transfer coefficient for a screen or felt wicked tube is similar to that of the bare tube and also not very different from that for the evaporating surface.

Marto and Mosteller(2.31) used a horizontal tube surrounded by four layers of 100 mesh stainless steel screen and water as the working fluid.   The outer container was of glass so it was possible to observe the evaporating wick surface.   They measured the superheat as a function of heat flux. They found that the critical radius was .013 mm compared to the effective capillary radius of 0.6 mm.

There is some evidence that the critical heat flux for wicked surfaces may be greater than that for smooth surfaces, for example a report by Costello and Frea (2.32) suggests that the critical heat flux could be increased by about 20% over that for a smooth tube.

Reiss and Schreitzman (2.33) report very high values of critical heat flux for sodium in grooved heat pipes.   They report values from two to ten times the critical values reported by Balzieser (2.34) in the temperature region around 550°C.   The authors observed the grooves as dark stripes on the outer side of the heat pipe and concluded that evaporation was from the grooves only.   Their results are plotted in Fig. 2.32 and the heat flux is calculated on the assumption that evaporation is from the groove only.

Moss and Kelley (2.35) constructed a planar evaporator using a stainless steel wick $\frac{1}{4}$ inch thick and water as the working fluid.   The authors employed a neutron radiographic technique to measure the water thickness. In order to explain their results they proposed two models, in the first it was assumed that vaporisation occurs at the liquid-vapour interface, in the second model a vapour blanket was assumed to occur at the base of the wick.

Both the Moss and Kelley models can be used to explain most of the published data.   These models are discussed in a paper by Ferrell, Davis and Winston

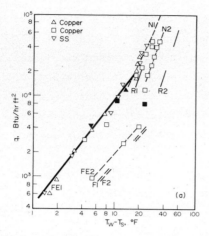

(a) Water boiling on the evaporator tube. Results are shown for three different tubes at atmospheric pressure. For one of the tubes results are shown for a pressure of 6 in. of mercury, these points beginning on line F_2 and ending on line N_2. Ordinate values multiplied by 3.16 give the flux in w / (m)^2. Shaded points indicate first visible nucleation.

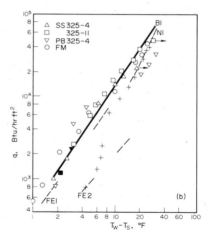

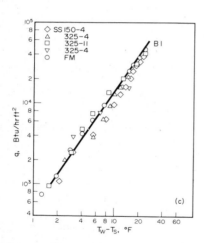

(b) Water boiling on the tube covered with a wick. The figure establishes Line B_1. The "plus" points originating at FE_2 are for the SS 150-4 wick at 6 in. mercury pressure, all other data at atmospheric pressure. Shaded points indicate first nucleation; an arrow is used for the plus points. Arrows at high fluxes indicate a progressive rise in temperature after the indicated point.

(c) Evaporation of water in the wick to its vapor. All data are for atmospheric pressure. Values of l vary from 1.5 to 4 in.

Fig. 2.31 Evaporation of water from smooth and wicked surfaces (2.30)

(2.36) who describe experiments aimed at differentiating between them.   The
two models are described by the authors as

    (1)   A layer of liquid filled wick adjacent to the heated
           surface with conduction across this layer and liquid
           vaporisation at the edge of the layer.   In this model
           liquid must be drawn into the surface adjacent by
           capillary forces.

    (2)   A thin layer of vapour filled wick adjacent to the
           heated surface with conduction across this layer to
           the saturated liquid within the wick.   In this model
           the liquid is vaporized at the edge of this layer and
           the resulting vapour finds its way out of the wick
           along the surface and through large pore size passages
           in the wick.   This model is analogous to conventional
           film boiling.

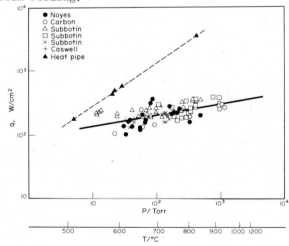

Fig. 2.32 Critical flux density for sodium pool boiling
compared to critical flux density in grooved sodium heat
pipes (2.33)

The apparatus used by Ferrell et al is shown in Fig. 2.33.

"This apparatus is designed to test flat, rectangular wicking materials
1.74 in. × 12 in.   The evaporator section consists of a 1.75 in. × 12 in.
heated surface.   A 3-1/2 in. diameter pressurized bellow presses one end of

the wicking material to this heated surface.   The other end of the wick
extends into a pool of liquid potassium which is maintained at the saturation
temperature by auxiliary heaters.   The adiabatic section is that portion of
the wick between the heated surface and the pool of metal.   The condenser
section is non-wicked and is situated in a reflux position to maintain a
constant pressure which is independent of the heat flux.

During the operation of the heat pipe, the graphite heater radiates thermal
energy to the back of a 304 stainless steel heater block.   This heat is
conducted to the wick where vaporization occurs.   Six thermocouples, three
each on two levels, give the steady-state linear gradient in the block.
This gradient is multiplied by the conductivity of the steel to give the
average heat flux and extrapolated to the surface of the block to give the
average surface temperature.   Six thermocouples mounted within 0.40 in. of
the surface detect any temperature excursions which might occur within the
wick.   Three additional thermocouples located in the liquid pool and vapour
space give the saturation temperatures in the apparatus.

An experimental run usually consisted of bringing the entire apparatus to
the saturation condition with the auxiliary heaters and allowing the potassium
to reflux on the wick for about 3 hours.   The power to the graphite heater
was then turned on and the temperatures recorded.   After steady state
conditions were reached, the temperatures were accurately determined using a
potentiometer.   A step change in the power was made and the procedure was
continued until either the critical heat flux or the maximum design heat
flux was exceeded.   The critical heat flux (heat flux at the wicking limit
of the heat pipe) was indicated by a temperature excursion in portions of the
wick due to a lack of coolant".

Experiments were carried out using a stainless steel felt wick FM1308 and
both water and potassium as the working fluids.   Though the results were
not entirely conclusive the author's believed that the second model is the
most likely mechanism.   Ferrell and Davis (2.37) report further work with
the same apparatus using potassium as the working fluid and both a stainless
steel felt wick FM1308 and a steel sintered powder wick Lamipore 7.4.   The
properties of these wicks are given in Table 2.5.

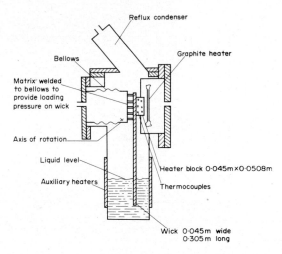

Fig. 2.33   Schematic of apparatus of Ferrell et al (2.36)

TABLE 2.5   DIMENSIONS AND PROPERTIES OF WICK MATERIALS

Material	Thickness m	Porosity	Permeability $m^2 \times 10^{10}$	Capillary rise m
FM1308	3.2×10^{-3}	0.58	0.55	0.26
Lamipore 7.4	1.5×10^{-3}	0.61	0.48	0.35

The data for the vertical heat pipe is different from the heat pipe in the horizontal positions for both types of wick.   Fig. 2.34 and 2.35 give results for the FM1308 wick in the vertical and horizontal positions.

For the vertical position the heat transfer coefficient for FM1308 increases with increasing flux becoming constant at a value 10.2 kW/m^2 °C.   For Lamipore 7.4 the coefficient decreases with increasing heat flux from 18.2 kW/m^2 °C to 14.8 kW/m^2 °C.

The effective thermal conductivities of the two wick structures have been calculated using the parallel model Equation 2.8.10 and the series model 2.8.11.   These results are given in Table 2.6 together with the heat transfer coefficients.

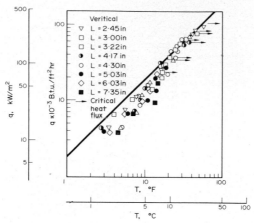

Fig. 2.34 Potassium boiling from vertical wick FM1308
(L_A is vertical height of wick above pool) (2.37)

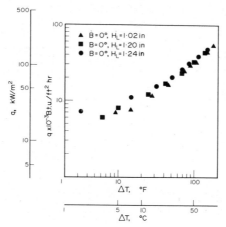

Fig. 2.35 Potassium boiling from horizontal wick FM1308
(2.37)

TABLE 2.6 HEAT TRANSFER COEFFICIENTS FOR POTASSIUM
IN VERTICAL HEAT PIPE (h in kW/m² °C)

Wick	Parallel model k_w	Series model k_w	Experimental result h
FM 1308	9.6	8.5	10.2
Lamipore 7.4	19	18	18.2 - 14.8

The agreement between the limits of the two models and the measured value
for FM1308 is close.   The agreement for Lamipore is good at low heat fluxes
and there was evidence for the development of a poor thermal contact during
the experiment which may explain the discrepancy at high flux values.

The heat transfer coefficients for the heat pipe in the horizontal position
were much lower than for the vertical case.   (1.1 kW/m^2 °C to 5 kW/m^2 °C)
and similar to results obtained for a bare horizontal surface with no
boiling.   It is suggested that in the horizontal case further temperature
drop occurs by conduction through an excess liquid layer above the wick
surface (this would not arise in normal heat pipe operation).

In the case of water the heat transfer coefficient is 11.3 kW/m^2 °C for both
the vertical and horizontal cases, Fig. 2.36.

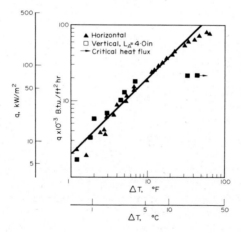

Fig. 2.36   Water boiling from FM1308 wick.   Data for
wick vertical and horizontal (2.37)

The authors conclude that the heat transfer mechanism is different for
liquid metals from the mechanism which applies to water and other non-
metallic fluids.   In the case of non-metallic fluids they suggest that the
vaporization process occurs within the porous media.   This vaporization is
probably initiated by inert gas trapped in the porous media or nucleation on
active sites on the heated surface.   Once initiated the vapour phase spreads
out to form a stable layer.

The data for water Fig. 2.36 shows that high fluxes are drawn for quite low
values of temperature difference whereas much larger values of temperature
difference might be expected if the mechanism is one of conventional film
boiling.   The mechanism may be of nucleate boiling at activation sites on
the heating surface and wick adjacent to it.   In view of the uncertainty of
the nature and position of the vapour origin more work is required.

It is difficult to initiate bubbles in liquid metals and the experimental
results of ref. 2.38 strongly suggest that for these fluids the wick is
saturated and that vaporisation occurs at the liquid surface on the outside
of the wick.   Hence for liquid metals the heat transfer coefficient can be
accurately predicted from Equations 2.8.10 and 2.8.11.

One limit to the radial heat transfer from the evaporator will be set by
'wicking', that is when capillary forces are unable to feed sufficient fluid
into the evaporator.

The limiting flux q_{crit} will be given by the expression

$$q_{crit} = \frac{\dot{m} \times L}{\text{area of evaporator}}$$

where the mass flow in is related to the pressure head ΔP by an expression
such as Equation 2.4.5.

Ferrell et al (2.39) have obtained such a relation for a planar surface and
a homogeneous wick.

$$q_{crit} = \frac{g(h_{co}\rho_{lo}\frac{\sigma_l}{\sigma_{lo}} - \rho_l l\sin\phi)}{\frac{l_e\mu_l}{L\rho_l kd}\left[\frac{l_e}{2} + l_a\right]} \qquad \ldots 2.8.7$$

where h_{co} is the capillary height of the fluid in the wick measured
          at a reference temperature

          σ_{lo}, ρ_{lo} are the fluid surface tension and density measured at
          the same reference temperature

σ_ℓ, ρ_ℓ, μ_ℓ are the fluid surface tension, density and viscosity measured at the operating temperature

All other symbols have their usual significance.

Ferrell and Davis (2.36) extended their equation by including a correction for thermal expansion of the wick.

$$q_{crit} = \frac{g\left[h_{co}\, \rho_{\ell o}\, \dfrac{\sigma_\ell}{\sigma_{\ell o}} - \rho_\ell \ell \, \sin\beta \, (1 + \alpha\Delta T)\right]}{\dfrac{\rho_e \mu_\ell}{L\rho_\ell \, Kd(1+\alpha\Delta T)} \left[\dfrac{\ell_e}{2} + \ell_a\right]} \qquad \ldots \; 2.8.8$$

where α is the coefficient of linear expansion of the wick

   ΔT is the difference in temperature between the operating
         and reference temperature

Fig. 2.37 shows a comparison of measured values of q_{crit} against predicted values from Equation 2.8.8 for both water and potassium. The equation successfully predicts heat flux limits for potassium up to a value of 315 kW/m^2.   It is also in good agreement for water up to 130 kW/m^2.   Above this value the experimental values fall below the values predicted by Equation 2.8.8 showing that another mechanism is limiting the flux.   The limiting factor is probably due to difficulty experienced by water vapour in leaving the heated surface through the wick.   The reduction in heat flux below the predicted value for water further supports the view that for non-metallic fluids vaporization occurs within the wick.

More recent work on radial heat flux measurements and observations of vapour/ liquid proportions and nucleation within wicks has resulted in more data showing that nucleation is not detrimental to heat pipe performance. Using water as the working fluid, this data is compared with some earlier results in Section 2.9.5 and the more important observations are also presented. Data on radial heat fluxes measured in heat pipes using a wide variety of other working fluids is given in Chapter 3.

The gravity-assisted heat pipe, which differs from the simple thermal syphon by retaining the wick, is also of interest from the point of view of radial heat fluxes, and is described in Section 2.10.

2.8.5 Liquid vapour interface temperature drop. Consider a liquid surface,
there will be a continuous flux of molecules leaving the surface by
evaporation. If the liquid is in equilibrium with the vapour above its
surface an equal flux of molecules will return to the liquid from the vapour
and there will be no net loss or gain of mass. However when a surface is
losing mass by evaporation clearly the vapour pressure and hence temperature
of the vapour above the surface must be less than the equilibrium value. In
the same way for net condensation to occur the vapour pressure and temperature
must be higher than the equilibrium value.

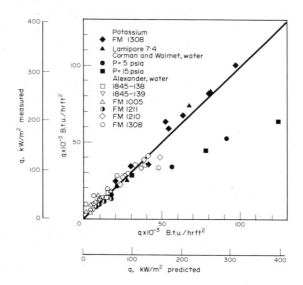

Fig. 2.37 Comparison of measured and predicted burnout
           for water and potassium (2.37)

The magnitude of the temperature drop can be estimated as follows.

The average velocity V_{av} in a vapour at temperature T_v and having molecular mass m is given by kinetic theory as

$$V_{av} = \sqrt{\frac{8\,kT_v}{\pi\,m}}$$

where k is Boltzmann's constant.

The average flow of molecules in any given direction is $\frac{nV_{av}}{4}$ /unit area. and the corresponding flow of heat/unit area

$$\frac{mLnV_{av}}{4}$$

where n is the number of molecules per unit volume and L the latent heat

Also for a perfect gas

$$\text{Pressure } P_v = nkT_v$$

$$\text{Hence heat flux} \quad \subseteq \quad P_v L \sqrt{\frac{m}{2\pi kT_v}}$$

This is the heat flux to the liquid surface, the flow of heat away from the surface will be given by:

$$P_\ell L \sqrt{\frac{m}{2\pi kT_\ell}}$$

Hence the net heat flux from the surface is

$$q = (P_\ell - P_v)L \sqrt{\frac{m}{2\pi kT_s}}$$

$$= \frac{P_\ell - P_v}{\sqrt{2\pi RT_s}} \qquad\qquad \dots 2.8.9(a)$$

where we have set $T_s \simeq T_v \simeq T_\ell$.

Some values of $\dfrac{q}{P_\ell - P_v}$ for liquids near their boiling points are given in Table 2.7.

By substituting Equation 2.8.1, 2.8.9(a) may be written:

$$q = \frac{\Delta T L^2 P}{(2\pi R T_s)^{\frac{1}{2}}} \times \frac{1}{R T_s^2} \qquad \qquad \ldots 2.8.9(b)$$

TABLE 2.7 (2.40)

Fluid	T_b K	$q/P_\ell - P_v$ kW/cm^2 atm
Lithium	1613	55
Zinc	1180	18
Sodium	1156	39
Water	373	21.5
Ethanol	351	13.5
Ammonia	238	15.2

2.8.6 Wick thermal conductivity.   The effective conductivity of the wick saturated with the working fluid is required for calculating the thermal resistance of the wick at the condenser region and also, under conditions of evaporation when the evaporation is from the surface for the evaporator region also.

Two models are used in the literature (see also Ch. 3, Section 3.3).

(i) Parallel case.   Here it is assumed that the wick and working fluid are effectively in parallel.

If k_ℓ is the thermal conductivity of the working fluid and k_s is the thermal conductivity of the wick material

and ε = Voidage fraction = $\dfrac{\text{Volume of working fluid in wick}}{\text{Total volume of wick}}$

$$k_w = (1 - \epsilon) k_s + \epsilon k_\ell \qquad \dots 2.8.10$$

(ii) Series case. If the two materials are assumed
to be in series

$$k_w = \frac{1}{\frac{1-\epsilon}{k_s} + \frac{\epsilon}{k_\ell}} \qquad \dots 2.8.11$$

Convection currents in the wick will tend to increase
the effective thermal conductivity.

2.8.7 Heat transfer in the condenser.

Vapour will condense on the liquid
surface in the condenser, the mechanism is similar to that discussed in
Section 2.8.5 on the mechanism of surface evaporation and there will be a
small temperature drop and hence thermal resistance. Further temperature
drops will occur in the liquid film and in the saturated wick and in the
heat pipe envelope.

The vapour liquid temperature drop may be calculated from Equations 2.8.10
and 2.8.11.

Condensation can occur in two forms, either by the condensing vapour forming
a continuous liquid surface or by forming a large number of drops. The
former, film condensation occurs in heat pipes and will be discussed here.
Condensation is seriously affected by the presence of a non-condensable gas.
However in the heat pipe vapour pumping will cause such gas to be concentrated
at the end of the condenser. This part of the condenser will be effectively
shut off and this effect is the basis of the gas buffered heat pipe.

The temperature drop through the saturated wick may be treated in the same
manner as at the evaporator.

Film condensation - Nusselt theory.

The first analysis of film condensation
was due to Nusselt and is given in the standard text books (2.15)(2.16).
The theory considers condensation onto a vertical surface and the resulting
condensed liquid film flows down the surface under the action gravity and
is assumed to be laminar.

Viscous shear between the vapour and liquid phases is neglected.   The mass
flow increases with distance from the top and the flow profile is shown in
Fig. 2.38.

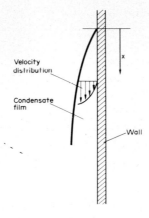

Fig. 2.38 Film condensation on a vertical surface

The average heat transfer coefficient h over a distance x from the top is
given by

$$h = 0.943 \left[\frac{L\rho_\ell^2 \, g \, k_\ell^3}{x\mu_\ell (T_s - T_w)} \right]^{\frac{1}{4}} \qquad \dots 2.8.12$$

where $T_s - T_w$ is the difference in temperature across the film

The Nusselt theory has been extended and applied to condensation in rotating
heat pipes, see Chap. 5, Section 5.7.

2.8.8 Total temperature drop.   Fig. 2.25 shows the components of the total
temperature drop along a heat pipe and the equivalent thermal resistances.

R_1 and R_9 are the normal heat transfer resistances for heating a solid
surface and are calculated in the usual way.

R_2 and R_8 represent the thermal resistance of the heat pipe wall.

R_3 and R_7 take account of the thermal resistance of the wick structure and
include any temperature difference between the wall and the liquid together

with conduction through the saturated wick.   From the discussion in
Section 2.8.4 it is seen that the calculation of R_3 is difficult if boiling
occurs, an upper limit may be obtained from Equations 2.8.10 and 2.8.11.
R_7 is made up principally from the saturated wick but if there is
appreciable excess liquid then a correction must be added (Section 2.8.7).

R_4 and R_6 are the thermal resistances corresponding to the vapour liquid
surfaces.   They may be calculated from Equation 2.8.7b but can usually be
neglected.

R_5 is due to the temperature drop ΔT_5 along the vapour column. ΔT_5 is
related to the vapour pressure drop ΔP_v by the Clapeyron equation 2.8.1

$$\frac{dP}{dT} = \frac{L}{TV_v}$$

or combining with the gas equation

$$\frac{dP}{dT} = \frac{LP}{RT^2}$$

Hence $\quad \Delta T_5 = \frac{RT^2}{LP} \Delta P_v$ $\qquad\qquad$ ... 2.8.12

where $\quad \Delta P_v$ is obtained from Section 2.5.

$\qquad \Delta T_5$ can usually be neglected.

R_s is the thermal resistance of the heat pipe structure, it can normally be
neglected but may be important in the start up of gas-buffered heat pipes.

In order to provide an indication of the relative magnitude of the various
thermal resistances Table 2.8 (from Asselman and Green (2.40)) lists some
approximate values/cm^2 of a water heat pipe.

TABLE 2.8

Resistance	$^\circ$C/w
R_1	$10^3 - 10$
R_2	10^{-1}
R_3	10
R_4	10^{-5}
R_5	10^{-8}
R_6	10^{-5}
R_7	10
R_8	10^{-1}
R_9	$10^3 - 10$

Formulae for the calculation of thermal resistance are given in Table 2.9.

TABLE 2.9   EQUIVALENT THERMAL RESISTANCES

Term	Defining relation	Thermal resistance	Comment
1	$Q_1 = h_e A_e \Delta T_1$	$R_1 = \dfrac{1}{h_e A_e}$	
2	Plane geometry $$q_e = \frac{k \Delta T_2}{t}$$ Cylindrical geometry $$q_e = \frac{2\pi k \Delta T_2}{\log_e \frac{r_2}{r_1}}$$	$R_2 = \dfrac{\Delta T_2}{A_e q_e}$	

3	Equation 2.8.10 or 2.8.11	$R_3 = \dfrac{d}{k_w A_e}$	Correct for liquid metals. Gives upper limit for non-metallics. See Section 2.8.4
4	$q_e = \dfrac{L^2 P_v}{(2\pi RT)^{\frac{1}{2}} RT^2}$	$R_4 = \dfrac{RT^2 (2\pi RT)^{\frac{1}{2}}}{L^2 P_v A_e}$	Usually can be neglected
5	$\Delta T_5 = \dfrac{RT^2 \Delta P_v}{LP_v}$ with ΔP_v from Section 2.5	$R_5 = \dfrac{RT^2 \Delta P_v}{QLP_v}$	Usually can be neglected
6	$q_c = \dfrac{\Delta T_6 L^2 P_v}{(2\pi RT)^{\frac{1}{2}} RT^2}$	$R_6 = \dfrac{RT^2 (2\pi RT)^{\frac{1}{2}}}{L^2 P_v A_c}$	Usually can be neglected
7	Equation 2.8.8 or 2.8.9	$R_7 = \dfrac{d}{k_w A_c}$	If excess working fluid present a correction should be made for the additional resistance
8	Plane geometry $q_c = \dfrac{k\Delta T_2}{t}$ Cylindrical geometry $q_c = \dfrac{2\pi k\Delta T_2}{\log_e \dfrac{r_2}{r_1}}$	$R_8 = \dfrac{\Delta T_8}{A_c q_c}$	For thin walled cylinders $\log_e \dfrac{r_2}{r_1} \simeq \dfrac{d}{r}$

| 9 | $Q_i = h_c A_c \Delta T_9$ | | $R_9 = \dfrac{1}{h_c A_c}$ | |

Notes   Resistance $= \dfrac{\text{Temperature difference}}{\text{Total heat flux}}$

Total heat flux $Q_i = Q + Q_s$

Flux density $\quad q = \dfrac{Q}{A}$

Q_i, Q and Q_s are defined in Fig. 2.25.

ΔT_1 to ΔT_9, R_1 to R_9, and R_s are also defined in this figure.

Other symbols used in the table are

A_e and A_c the evaporator and condenser wall areas

h_e and h_c heat transfer coefficients at the evaporator and condenser
        outer surfaces

q_e and q_c heat flux through the evaporator and condenser walls

k thermal conductivity of the heat pipe wall

t thickness of the heat pipe wall

r_1, r_2 inner and outer radius of cylindrical heat pipe

d wick thickness

k_w effective wick thermal conductivity

P_v vapour pressure

L latent heat

R Gas constant for vapour $R = \dfrac{Ro}{M}$

T Absolute temperature of vapour

ΔP_v total vapour pressure drop in pipe

2.9 Limits to Heat Transport

Upper limits to the heat transport capability of a heat pipe may be set
by one or more factors.   These limits are illustrated in Fig. 2.1.

2.9.1 Viscous limit.   At low temperatures viscous forces are dominant

in the vapour flow down the pipe.   Busse has shown that the axial heat
flux rapidly increases as the pressure in the condenser is reduced, the
maximum heat flux occurring when the pressure is reduced to zero.   Busse
carried out a two dimensional analysis, finding that the radial velocity
component had a significant effect, he derived the following equation

$$q = \frac{r_v^2 L \, \rho_v P_v}{16 \mu_v \ell_{eff}} \qquad\qquad \ldots \; 2.9.1$$

where P_v and ρ_v refer to the evaporator end of the pipe.

This equation agrees well with published data.

2.9.2 Sonic limit.   At a somewhat higher temperatures choking at the

evaporator exit may limit the total power handling capability of the pipe.
This problem was discussed in Section 2.5.

The sonic limit is given by:

$$q = 0.474 \, L \, (\rho_v P_v)^{\frac{1}{2}} \qquad\qquad \ldots \; 2.9.2$$

There is good agreement between this formula and experimental results.
Fig.2.39 due to Busse plots the temperature at which the sonic limit is
equal to the viscous limit as a function of ℓ_{eff}/d^2 for some alkali metals,
where $d = 2r_v$ the vapour space diameter.

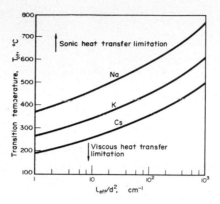

Fig. 2.39   Transition from viscous to sonic limitation

2.9.3 Entrainment limit.   This was discussed in Section 2.7.   Equation 2.7.3 gives the entrainment limiting flux

$$q = \sqrt{\frac{2\pi \rho_v L^2 \sigma_\ell}{Z}} \qquad \ldots 2.7.3$$

Kemme's experiments suggest a rough correlation of this failure mode with the centre-to-centre wire spacing for Z, and that a very fine screen will suppress entrainment.

2.9.4 Capillary limit, (wicking limit).   In order for the heat pipe to operate, Equation 2.1.1 must be satisfied, namely

$$\Delta P_{c_{max}} \geqslant \Delta P_\ell + \Delta P_v + \Delta P_g$$

Expressions to enable these quantities to be calculated are given in Sections 2.2, 2.3, 2.4, 2.5 and 2.6.

An expression for the maximum flow rate $\dot{m}_{max}$ may readily be obtained if we assume

   (i)   The liquid properties do not vary along the pipe

(ii)   The wick is uniform along the pipe

(iii)   The pressure drop due to vapour flow can be neglected.

$$\dot{m}_{max} = \left[\frac{\rho_\ell \sigma_\ell}{\mu_\ell}\right]\left[\frac{KA}{\ell}\right]\left[\frac{2}{r_e} - \frac{\rho_\ell g \ell}{\sigma_\ell} \sin\phi\right] \qquad \ldots 2.9.3$$

and the corresponding heat transport Q_{max}

$$Q_{max} = \dot{m}_{max} L$$

$$= \left[\frac{\rho_\ell \sigma_\ell L}{\mu_\ell}\right]\left[\frac{KA}{\ell}\right]\left[\frac{2}{r_e} - \frac{\rho_\ell g \ell}{\sigma_\ell} \sin\phi\right] \qquad \ldots 2.9.4$$

The group $\dfrac{\rho_\ell \sigma_\ell L}{\mu_\ell}$ depends only on the properties of the working fluid and is known as the figure of merit M.

It is plotted vs temperature for a number of working fluids in Fig. 2.2 $\dfrac{KA}{r_e}$ specifies the wick geometrical properties.   The experimental limits at the higher temperatures in Fig. 2.2 are believed to represent a wicking limit.

2.9.5 Burnout.   Burnout will occur at the evaporator at high radial fluxes. A similar limit on peak radial flux will also occur at the condenser.   These limits are discussed in Section 2.8.   At the evaporator Equation 2.8.8 gives a limit which must be satisfied for a homogeneous wick.   This equation, which represents a wicking limit, is shown in Section 2.8.4 to apply to potassium up to 315 kW/m^2 and probably is applicable at higher values for potassium and other liquid metals.   For water and other non metallics vapour production in the wick becomes important at lower flux densities (130 kW/m^2 for water) and there are no simple correlations.   For these fluids experimental data in Table 3.2 should be used as an indication of flux densities which can be achieved.

As discussed in Section 2.8.4, boiling in wicks is a topic of considerable interest, and since the first edition of this book was published, extra data,

particularly associated with water as the working fluid, has become available.
Tien (2.43) has summarized the problem of nucleation within the wick, and the
arguments for possible performance enhancement in the presence of nucleation.
Because of phase equilibrium at the interface, liquid within the wick at the
evaporator is always superheated to some degree, but it is difficult to
specify the degree of superheat needed to initiate bubbles in the wick. While
boiling is often taken as a limiting feature in wicks, possibly upsetting the
capillary action, Cornwell and Nair (2.44) found that nucleation reduces the
radial temperature difference and increases heat pipe conductance. One factor
in support of this is the increased thermal conductivity of liquid saturated
wicks in the evaporator section, compared with that in the condenser.

Tien believed that the boiling limit will become effective only when the
bubbles generated within the wick become trapped there, forming a vapour
blanket. Thus, some enhancement of radial heat transfer in the evaporator
section can be obtained by nucleate boiling effects.

It is useful to compare the measurements of radial heat flux as a function
of the degree of superheat (in terms of the difference between wall temperature T_w and saturation temperature T_{sat}) for a number of surfaces and wick
forms which have been the subject of recent studies. These are shown in
Table 2.10.

Abhat and Seban (2.30), in discussing their results as shown in Table 2.10,
stated that for every pool depth, there was a departure from the performance
recorded involving a slow increase in the evaporator temperature. In order
to guarantee indefinite operation of the heat pipe without dry-out, the
authors therefore suggested that the operating flux should be considerably
less than that shown in the Table. They also found that their measurements
of radial flux as a function of T_w - T_{sat} were similar, regardless of whether
the surface was flat, contained mesh, or a felt. This is partly borne out
by the results of Wiebe and Judd (2.45), also given in Table 2.10. Costello
and Frea, however, dispute this (2.32).

The work of Cornwell and Nair (2.44), which is currently being extended, gives
results similar to those of Abhat and Seban. Some of Cornwell and Nair's
results are presented in Fig. 2.40 and compared with theory. Cornwell and
Nair found that some indication of the q/ΔT curve for boiling within a wick

TABLE 2.10   MEASURED RADIAL HEAT FLUXES IN WICKS

Reference	Wick	Working Fluid	Superheat (^{o}C)	Flux (W/cm^2)
Wiebe & Judd (2.45)	Horizontal flat surface	Water	3	0.4
			6	1.2
			11	8
			17	20
Abhat & Seban (2.30)	Meshes and Felts	Water	5	1.6
			6	5
			11	12
			17	20
Cornwell & Nair (2.44)	Foam	Water	5	2
			10	10
			20	18
	Mesh (100)	Water	5	7
			10	9
			20	13
Abhat & Nguyenchi (2.50)	Mesh*	Water	6	1
			10	8

*Gravity assisted

could be obtained by assuming that boiling occurs only on the liquid covered area of the heating surface (measured from observations) and that q based on this area may be expressed by the nucleate boiling correlation:

$$q = (K\Delta T)^a$$

where   K = Constant
        a = index in range 3 - 6

The analysis is restricted to situations where the vapour formed in the wick escapes through the wick surface, and not out of the sides or through grooves in the heating surface.

The attempt to relate the flux to the area covered by liquid is a completely different approach to that of Nishikawa and Fujita (2.56), who have investigated the effect of bubble population density on the heat flux limitations and the degree of superheat. While this work is restricted at present to flat surfaces the authors' new nucleation factor may become relevant in studies where the wick contains channels or consists solely of grooves, possibly flooded, in the heat pipe wall.

HP—D

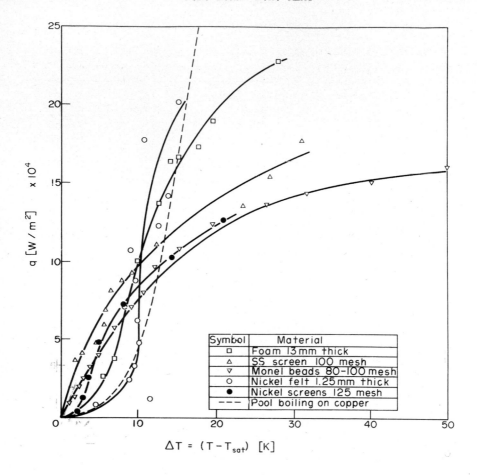

Fig. 2.40 Variation of heat flux with superheat for water
           in porous media (2.44).

The work of Saaski (2.46) on an inverted meniscus wick is of considerable
interest. This concept, illustrated in Fig. 2.41, embodies the high heat
transfer coefficient of the circumferential groove while retaining the circumferential fluid transport capability of a thick sinter or wire mesh wick.
With ammonia, heat transfer coefficients in the range 2 to 2.7 W/m^2K were
measured at radial heat flux densities of 20 W/cm^2. These values were significantly higher than those obtained for other non-boiling evaporative surfaces.
Saaski suggested that the heat transfer enhancement may be due to film tur-
bulence generated by vapour shear, or surface tension-driven convection. He
contemplated an increase in groove density as one way of increasing the heat

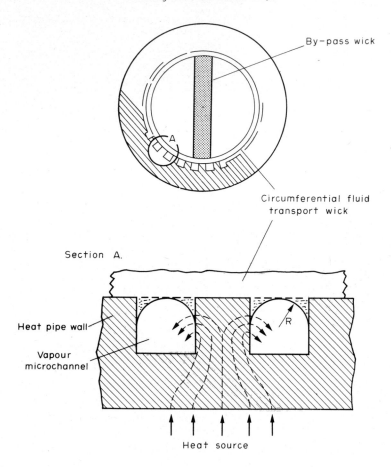

Fig. 2.41   High performance inverted meniscus hybrid wick
            proposed by Saaski (2.46)

transfer coefficient, as first results have indicated a direct relationship
between groove density and heat transfer coefficient. Vapour resulting from
evaporation at the inverted meniscus interface flows along to open ends of the
grooves, where it enters the central vapour core.

The main theoretical requirement for inverted meniscus operation, according
to Saaski, is the maintenance of a sufficiently low vapour pressure drop in
each channel of the wick. Recession of the inverted meniscus, which is the
primary means of high evaporative heat transfer, reduces capability considerably. The equation given below describing the heat flux capability contains
a term (j) which is a function of vapour microchannel pressure drop, being

defined as the ratio of this pressure drop to the maximum capillary priming
potential.

$$\text{Thus} \quad j \ = \ \frac{\Delta P_v}{2\sigma/r_c}$$

The heat flux q_{max} is defined by:

$$q_{max} \ = \ \frac{jN^2X^4}{8\pi}\left(\frac{\rho_v L\sigma}{\mu_v M}\right)\frac{1}{N^3 d_g^2 r_c} \qquad \qquad \cdots 2.9.5$$

$$\text{where } X \ = \ Cos\left(\frac{\psi}{2}\right) \Big/ \left[1 + Sin\left(\frac{\psi}{2}\right)\right]$$

ψ = groove angle (degrees)

M = molecular weight of working fluid

N = number of grooves/cm

d_g = heat pipe diameter at inner groove radius

(q_{max} is given in W/cm^2 by equation 2.9.5)

Feldman and Berger (2.47) carried out a theoretical analysis to determine
surfaces having potential in a high heat flux water heat pipe evaporator.
Following a survey of the literature, circumferential grooves were chosen,
these being of rectangular and triangular geometry. The model proposed that:

(i)   At heat fluxes below nucleate boiling, conduction is the main
      mode of heat transfer and vaporization occurs at the liquid/
      vapour interface without affecting capillary action.

(ii) As nucleation progresses, the bubbles are readily expelled
      from the liquid in the grooves, with local turbulence, and
      convection becomes the main mode of heat transfer.

(iii) At a critical heat flux, nucleation sites will combine
      forming a vapour blanket, or the groove will dry out.
Both result in a sharp increase in evaporator temperature.

It is claimed that film coefficients measured by other workers supported the
computer model predictions, and Feldman stated that the model indicated that
evaporator film coefficients as high as 8 W/cm^2 K could be obtained with
water as the working fluid. Using triangular grooves, it was suggested that
radial fluxes of up to 150 W/cm^2 could be tolerated. However, such a

suggestion appears largely hypothetical, bearing in mind the measured values
given in Table 2.10.

Winston, Ferrell and Davis (2.54) reported further progress on the prediction
of maximum evaporator heat fluxes, including a useful alternative to equation
2.8.7. Based on modification of an equation first developed by Johnson (2.55)
the new relationship given below takes into account vapour friction in the
wick evaporator section:

$$q_{crit} = \frac{g \left[h_{co} \, \rho_{\ell o} \, \dfrac{\sigma_{\ell}}{\alpha_{1o}} - \rho_{\ell} . (\ell_a + \ell_e) \, Sin \, \phi \right]}{\dfrac{\ell_e \, \mu_{\ell}}{L\rho_{\ell} \, K\varepsilon (r_{\omega} - r_v)} \left[\dfrac{\ell_e}{2} + \ell_a \right] + \dfrac{\mu_v}{K\rho_v} . \dfrac{(r_{\omega} - r_v)}{(\varepsilon - \varepsilon_{\ell}) L}} \qquad \ldots \, 2.9.6$$

where nomenclature is as defined in Section 2.8.4 The solution of the equation was accomplished by arbitrarily varying the porosity for liquid flow,
ε_{ℓ} from zero to a maximum given by the wick porosity ε. The portion of the
wick volume not occupied by liquid was assumed to be filled with vapour.

Equation 2.9.6 has been used to predict critical heat fluxes, and a comparison
with measured values for water has shown closer agreement than that obtained
using equation 2.8.7 or 2.8.8, although the validity of these two latter
equations is not disputed for liquid metal heat pipes.

2.10 Gravity Assisted Heat Pipes

The use of gravity-assisted heat pipes as opposed to reliance on simple
thermal syphons, has only received detailed attention in the past three to
four years. The main areas of study have centred on the need to optimise
the fluid inventory, and to develop 'wicks' which will minimise entrainment.

Some interesting results on gravity-assisted heat pipes were presented at
the 2nd International Heat Pipe Conference. Deverall and Keddy (2.48) used
helical arteries, in conjunction with meshes for evaporator liquid distribution, and obtained relatively high axial fluxes, albeit with sodium and
potassium as the working fluids. This, together with the work of Kemme
(2.49), led to the recommendation that more effort be devoted to the vapour
flow limitations in gravity-assisted heat pipes. To date this work has been

restricted to heat pipes having liquid metal working fluids, and the significance in water heat pipes remains questionable.

However, for liquid metal heat pipes, Kemme discusses in some detail a number
of vapour flow limitations in heat pipes operating with gravity assistance. As
well as the pressure gradient limit discussed above, Kemme presents equations
for the viscous limit, described by Busse as:

$$q_v = \frac{A_v \, D^2 \, L}{64 \, \mu_v \, \ell_{eff}} \, \rho_{ve} \, P_{ve} \qquad \qquad \ldots 2.10.1$$

where A_v is the vapour space cross-sectional area, D is the vapour passage
diameter and suffix e denotes conditions in the evaporator. (See also Section
2.9.1). Kemme also suggested a modified entrainment limitation to cater for
additional buoyancy forces during vertical operation:

$$Q_{max} = A_v \, L \left[\frac{\rho_v}{A} \left(\frac{2\pi\gamma}{\lambda} + \rho \, gD \right) \right]^{\frac{1}{2}} \qquad \ldots 2.10.2$$

where λ is the characteristic dimension of the liquid/vapour interface, and
was calculated as d_W plus the distance between the wires of the mesh wick used.
$\rho \, gD$ is the buoyancy force term.

Abhat and Nguyenchi (2.50) report work carried out at IKE, Stuttgart, on gravity-assisted copper/water heat pipes, retaining a simple mesh wick located
against the heat pipe wall. Tests were carried out with heat pipes at a number
of angles to the horizontal, retaining gravity assistance, with fluid invent-
ories of up to 5 times that required to completely saturate the wick. Thus a
pool of liquid was generally present in at least part of the evaporator section. The basis of the analysis carried out by Abhat and Nguyenchi to compare
their experimental results with theory was the model proposed by Kaser (2.51)
which assumed a liquid puddle in the heat pipe varying along the evaporator
length from zero at the end of the heat pipes to a maximum at the evaporator
exit. Results obtained by varying the operating temperature, tilt angle and
working fluid inventory (see Fig 2.42) were compared with Kaser's model, which
was, however, found to be inadequate. Kaser believed that the limiting value

of heat transport occurred when the puddle commenced receding from the end of the evaporator. However, the results of Abhat and Nguyenchi indicate that the performance is limited by nucleate boiling in the puddle. While these results are of interest, much more work is needed in this area.

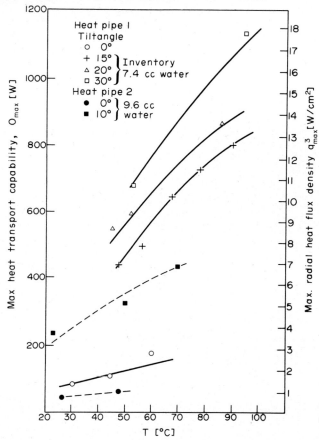

Fig. 2.42   The effect of temperature on heat pipe
performance for various tilt angles (gravity
assisted angle measured from the horizontal)
(2.50)

One contribution to the performance of gravity-assisted heat pipes and thermal syphons was reported in 1975. Strel'tsov (2.52) carried out a theoretical and experimental study to determine the optimum quantity of fluid to use in gravity-assisted units. Without quantifying the heat fluxes involved, he derived expressions to permit determination of the fluid inventory. Of particular interest was his observation that film evaporation, rather than nucleation, occurred under all conditions in the evaporator section, using water and seve-

ral organic fluids, this seeming to contradict the findings of Abhat and Nguyenchi, assuming that Strel'tsov achieved limiting heat fluxes.
Strel'tsov derived the following expression for the optimum fluid inventory
for a vertical heat pipe:

$$G = (0.8 \; \ell_c \, \ell_a + 0.8 \; \ell_e) \sqrt[3]{\frac{3Q\mu_1\rho_1\pi^2D^2}{Lg}} \qquad \ldots 2.10.3$$

where Q is the heat transport (Watts).

For a given heat pipe design and assumed temperature level, the dependence of
the optimum quantity of the working fluid is given by the expression:

$$G = K \sqrt[3]{Q} \qquad \ldots 2.10.4$$

where K is a function of the particular heat pipe under consideration. The
predicted performance of a heat pipe using methanol as the working fluid at a
vapour temperature of $55^\circ C$ is compared with measured values in Fig. 2.43

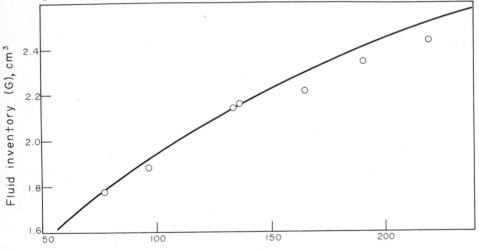

Fig 2.43. The measured effect of fluid inventory
on the performance of a methanol heat
pipe, and a comparison with results
predicted by Eqn.2.10.3 (2.52).

The use of arteries in conjunction with grooved evaporator and condenser surfaces has been proposed by Vasiliev and Kiselyov (2.53). The artery is used
to transfer condensate between the condenser and evaporator, thus providing an
entrainment-free path, while triangular grooves in the evaporator and condenser wall are used for distribution and collection of the working fluid. It is

claimed that this design has a higher effective thermal conductivity than a simple thermal syphon, particularly at high heat fluxes.

The equation developed for the maximum heat flux of such a heat pipe is:

$$q_{max} \ (W/m^2) = 3.26 \times 10^{-2} \ a \ Cos \ \frac{\psi}{2} \ Cos \ \theta. \ \theta(\Psi). \frac{\sigma\rho_\ell L}{\mu_\ell} \quad \ldots \ 2.10.5$$

where a = groove width

ψ = groove angle

θ = contact angle

$$f(\psi) = [\ tg \ \psi + c.tg \ \psi - \frac{(\frac{\pi}{2} - \psi)}{Cos^2 \ \psi} \ ]^3 \ Cos^2 \ \psi$$

(Fluid inventory is also discussed in Section 3.6).

REFERENCES

2.1   Cotter, T.P. Theory of heat pipes, LA-3246-MS, 26 March 1965.

2.2   Shaw, D.J. Introduction to colloid and surface chemistry. Pub. Butterworth, 2nd Ed. 1970.

2.3   Semenchenke, V.K. Surface phenomena in metals and alloys. Pergamon 1961.

2.4   Bohdansky, J., Schins, H.E.J. The surface tension of the alkali metals. J. Inorg. Nucl. Chem. 29 (1967) pp 2173 - 2179.

2.5   Bohdansky, J., Schins, H.E.J. The temperature dependence of liquid metals. J. Chem. Phys. Vol. 49, p. 2982, 1968.

2.6   Busse, C.A. Pressure drop in the vapour phase of long heat pipes. Thermionic Conversion Specialists Conference, Palo Alto, Calif., October 1967.

2.7   Grover, G.M., Kemme, J.E., Keddy, E.S. Advances in heat pipe technology. 2nd International Symposium on Thermionic Electrical Power Generation. Stresa. Italy, May 1968.

2.8   Ernst, D.M. Evaluation of theoretical heat pipe performance.
         Thermionic Specialist Conference, Palo Alto, Calif. October 1967.

2.9   Bankston, C.A., Smith, J.H. Incompressible laminar vapour flow in
         cylindrical heat pipes. ASME Paper No. 71 - WA/HT-15, 1971.

2.10   Rohani, A.R., Tien, C.L. Analysis of the effects of vapour pressure
         drop on heat pipe performance. Int. J. Heat Mass Trans. Vol. 17,
         pp 61 - 67.

2.11   Deverall, J.E., Kemme, J.E., Flarschuetz, L.W. Some limitations and
         start-up problems of heat pipes. LA -4518, November 1970.

2.12   Kemme, J.E. Ultimate heat pipe performance. I.E.E.E. Thermionic
         Specialist Conference. Framingham Mass., 1968.

2.13   Kemme, J.E. High performance heat pipes. I.E.E.E. Thermionic
         Specialist Conference. Palo Alto, Calif., October 1967.

2.14   Cheung, H. A critical review of heat pipe theory and applications.
         UCRL-50453, July 1968.

2.15   McAdams, W.H. Heat Transmission, 3rd Ed. McGraw-Hill, New York
         (1954).

2.16   Tong, L.S. Boiling heat transfer and two phase flow. John Wiley
         (1965).

2.17   Nukiyama, S. Maximum and minimum values of heat transmitted from
         metal to boiling water under atmospheric pressure. J. Soc. Mech. Eng.
         Japan, Vol. 37, p. 367 (1934).

2.18   Eckert, E.R. Gu and Drake, R.M. Heat and mass transfer. 2nd Ed.
         McGraw-Hill, New York (1959).

2.19   Schins, H.E.J. Comparative study of wick and pool boiling sodium
         systems. International Heat Pipe Conference. Stuttgart, 1973.

2.20   Hsu, Y.Y. On the size range of active nucleation cavities on a
         heating surface. J. of Heat Transfer, Trans A.S.M.E. August 1962.

2.21   Griffith, P. and Wallis, J.D. The role of surface conditions in
         nucleate boiling. Tech. Report No. 14, Contract N5-OR1-07894. M.I.T.
         Div. of Ind. Coop. December 1958.

2.22   Rohsenhow, W.M. A method of correlating heat transfer data for
         surface boiling of liquids. Trans. A.S.M.E., Vol 74, 1955.

2.23   Rohsenhow, W.M., Griffith, P. Correlation of maximum heat flux data
         for boiling of saturated liquids. A.S.M.E. - A.I.C.E. Heat Transfer
         Symposium, Louisville, Ky., 1955.

2.24   Caswell, B.F., Balzhieser, R.E. The critical heat flux for boiling
         metal systems. Chem. Eng. Prog. Symp. Series on Heat Transfer. Los
         Angeles, No. 64, Vol 62, 1966.

2.25   Subbotin, V.I. Heat transfer in boiling metals by natural convec-
         tion. USAEC - Tr - 7210, 1972.

2.26   Dwyer, O.E. On incipient boiling wall superheats in liquid metals,
         Int. J. Heat Mass Transfer, Vol 12, pp 1403 - 1419, 1969.

2.27   Philips, E.C., Hinderman, J.D. Determination of capillary properties
         useful in heat pipe design. A.S.M.E. - A.I.Ch.E. Heat Transfer Conf.
         Minneapolis, Minnesota, August 1967.

2.28   Ferrell, J.K., Alleavitch, J. Vaporisation heat transfer in capil-
       lary wick structures. Chem. Eng. Symp. Series. No 66, Vol. 02, 1970.

2.29   Corman, J.C., Welmet, C.E. Vaporisation from capillary wick struc-
       tures. A.S.M.E. - A.I.Ch.E. Heat Transfer Conference, Tulsa, Okla-
       homa. Paper 71-HT-35, August 1971.

2.30   Abhat, A., Seban, R.A. Boiling and evaporation from heat pipe wicks
       with water and acetone. Jour. Heat Transfer, August 1974.

2.31   Marto, P.S., Mosteller, W.L. Effect of nucleate boiling and the
       operation of low temperature heat pipes. A.S.M.E.-A.I.Ch.E. Heat
       Transfer Conference, Minneapolis, Minnesota, August 1969.

2.32   Costello, C.P., Frea, W.J. The roles of capillary wicking and sur-
       face deposits in attainment of high pool boiling burnout heat fluxes:
       A.I.Ch.E. Jour. 1964, No. 10 (3), p 393.

2.33   Reiss, F. et al. Pressure balance and maximum power density at the
       evaporator gained from heat pipe experiments. 2nd International Sym-
       posium on Thermionic Electrical Power Generation. Stresa, Italy,
       May 1968.

2.34   Balzhieser, R.E. et al. Investigation of liquid metal boiling heat
       transfer. Air Force Propulsion Lab. Wright Patterson, AFB/Ohio, 1966,
       AFAPL-TR-66-85.

2.35   Moss, R.A., Kelley, A.J. Neutron radiographic study of limiting
       heat pipe performance. Int. J. Heat Mass Transfer, Vol 13 (3),
       pp 491 - 502, 1970.

2.36   Ferrell, J.K., Davis, R., Winston, H. Vaporisation heat transfer in
       heat pipe wick materials. International Heat Pipe Conference, Stutt-
       gart, October 1973.

2.37   Davis, W.R., Ferrell, J.K. Evaporative heat transfer of liquid pot-
       assium in porous media. A.I.A.A./A.S.M.E. 1974. Thermophysics and
       Heat Transfer Conference, Boston, Mass. July 1974.

2.38   Alexandra, E.G. Structure property relationship in heat pipe wicking
       materials. PhD Thesis. N.C. State University, 1972.

2.39   Ferrell, J.K. et al. Vaporisation heat transfer in heat pipe wick
       materials. A.I.A.A. Thermophysics Conference, San Antonio, Texas,
       April 1972.

2.40   Asselman, G.A.A., Green, D.B. Heat Pipes. Philips Tech. Rev. Vol 33,
       pp 104 - 113, No. 4, 1973.

2.41   Busse, C.A. Theory of ultimate heat transfer limit of cylindrical
       heat pipes. Inst. J. Heat and Mass Transfer. Vol. 16, pp 169 - 186,
       1973.

2.42   Busse, C.A., Vinz, P. Axial heat transfer limits of cylindrical
       sodium heat pipes between 25 W/cm^2 and 15.5 kW/cm^2. International
       Heat Pipe Conference. Stuttgart, October 1973.

2.43   Tien, C.L. Fluid mechanics of heat pipes. Annual Review Fluid Mech.,
       Am. Inst. Physics, Vol. 7, pp. 167 - 185, 1975.

2.44   Cornwell, K. and Nair, B.G. Boiling in wicks. Proc. Heat Pipe Forum
       Meeting, Publ. as National Engineering Laboratory. Report No. 607,
       1976.

2.45   Wiebe, J.R. and Judd, R.L. Superheat layer thickness measurements
        in saturated and subcooled nucleate boiling. Trans. ASME, J. Heat
        Transfer, Paper 71-HT-43, 1971.

2.46   Saaski, E.W. Investigation of an inverted meniscus heat pipe wick
        concept. NASA Report CR-137724, 1975.

2.47   Feldman, K.T. and Berger, M.E. Analysis of a high heat flux water
        heat pipe evaporator. Tech. Report ME-62(73), ONR-012-2, U.S. Office
        of Naval Research, 1973.

2.48   Deverall, J.E. and Keddy, E.S. Helical wick structures for gravity-
        assisted heat pipes. Proc. 2nd Int. Heat Pipe Conference, Bologna;
        ESA Report SP-112, Vol. 1, 1976.

2.49   Kemme, J.E. Vapor flow considerations in conventional and gravity-
        assist heat pipes. Proc. 2nd Int. Heat Pipe Conference, Bologna; ESA
        Report SP 112, Vol. 1, 1976.

2.50   Abhat, A. and Nguyenchi, H. Investigation of performance of gravity
        assisted copper-water heat pipes. Proc. 2nd Int. Heat Pipe Conference,
        Bologna; ESA Report SP112, Vol. 1, 1976.

2.51   Kaser, R.V. Heat pipe operation in a gravity field with liquid pool
        pumping. Unpublished paper, McDonnell Douglas Corporation, July 1972.

2.52   Strel'tsov, A.I. Theoretical and experimental investigation of op-
        timum filling for heat pipes. Heat Transfer-Soviet Research, Vol. 7,
        No. 1, Jan/Feb 1975.

2.53   Vasiliev, L.L. and Kiselyov, V.G. Simplified analytical model of
        vertical arterial heat pipes. Proc. 5th Int. Heat Transfer Conference,
        Japan; Vol. 5, Paper HE 2.3, 1974

2.54   Winston, H.M., Ferrell, J.K. and Davis, R. The mechanism of heat
        transfer in the evaporator zone of the heat pipe. Proc. 2nd Int.
        Heat Pipe Conference, Bologna; ESA Report SP112, Vol. 1, 1976.

2.55   Johnson, J.R. and Ferrell, J.K. The mechanism of heat transfer in the
        evaporator zone of the heat pipe. ASME Space Techn. & Heat Transfer
        Conference, Los Angeles; Paper 70-HT/SpT-12, 1970.

2.56   Nishikawa, K. and Fujita, Y. Correlation of nucleate boiling heat
        transfer based on bubble population density. Int. J. Heat Mass
        Trans., Vol. 20, pp. 233-245, 1977.

Practical Design Considerations

The three basic components of a heat pipe are:

(i)   The working fluid

(ii)   The wick or capillary structure

(iii)   The container

In the selection of a suitable combination of the above, inevitably a number of conflicting factors may arise, and the principle bases for selection are discussed below:

3.1 The Working Fluid

A first consideration in the identification of a suitable working fluid is the operating vapour temperature range and a selection of fluids is shown in Table 3.1. Within the approximate temperature band several possible working fluids may exist, and a variety of characteristics must be examined in order to determine the most acceptable of these fluids for the application being considered. The prime requirements are:

(i)   Compatibility with wick and wall materials

(ii)   Good thermal stability

(iii)   Wettability of wick and wall materials

(iv)   Vapour pressures not too high or low over the operating temperature range

(v)   High latent heat

(vi)   High thermal conductivity

(vii)   Low liquid and vapour viscosities

(viii) High surface tension

(ix)   Acceptable freezing or pour point

TABLE 3.1 HEAT PIPE WORKING FLUIDS

(The useful operating temperature range is indicative only). Full properties of most of the below are given in Appendix 1.

Medium	Melting point (°C)	Boiling point at atmos.press. (°C)	Useful range (°C)
Helium	-272	-269	-271 - -269
Nitrogen	-210	-196	-203 - -160
Ammonia	-78	-33	-60 - 100
Freon 11	-111	24	-40 - 120
Pentane	-130	28	-20 - 120
Freon 113	-35	48	-10 - 100
Acetone	-95	57	0 - 120
Methanol	-98	64	10 - 130
Flutec PP2*	-50	76	10 - 160
Ethanol	-112	78	0 - 130
Heptane	-90	98	0 - 150
Water	0	100	30 - 200
Flutec PP9*	-70	160	0 - 225
Thermex	12	257	150 - 395
Mercury	-39	361	250 - 650
Caesium	29	670	450 - 900
Potassium	62	774	500 - 1000
Sodium	98	892	600 - 1200
Lithium	179	1340	1000 - 1800
Silver	960	2212	1800 - 2300

* Included for cases where electrical insulation is a requirement

The selection of the working fluid must also be based on thermodynamic considerations which are concerned with the various limitations to heat flow occuring within the heat pipe. These were also discussed in Chapter 2 and are the viscous, sonic, capillary, entrainment and nucleate boiling limitations.

Many of the problems associated with long life heat pipe operation are a direct consequence of material incompatibility. This involves all three components of the heat pipe, and is discussed fully later. One aspect peculiar to the working fluid, however, is the possibility of thermal degradation. With certain organic fluids it is necessary to keep the film temperature below a specified value to prevent the fluid breaking down into

different compounds. A good thermal stability is therefore a necessary
feature of the working fluid over its likely operating temperature range.

The surface of a liquid behaves like a stretched skin except that the tension
in the liquid surface is independent of surface area. All over the surface
area of a liquid there is a pull due to the attraction of the molecules tend-
ing to prevent their escape. This surface tension varies with temperature
and pressure, but the variation with pressure is frequently small.

The effective value of surface tension may be considerably altered by the
accumulation of foreign matter at the liquid/vapour liquid/liquid, or solid
surfaces. Prediction of surface tension is discussed in Chapter 2.

In heat pipe design a high value of surface tension is desirable in order to
enable the heat pipe to operate against gravity and to generate a high
capillary driving force.

In addition to high surface tension, it is necessary for the working fluid to
wet the wick and container material. That is the contact angle must be zero,
or at least very small.

The vapour pressure over the operating temperature range must be sufficiently
great to avoid high vapour velocities which tend to set up a large temperature
gradient, entrain the refluxing condensate in the counter current flow, or
cause flow instabilities associated with compressibility. However, the
pressure must not be too high because this will necessitate a thick-walled
container.

A high latent heat of vaporisation is desirable in order to transfer large
amounts of heat with a minimum fluid flow, and hence to maintain low pressure
drops within the heat pipe. The thermal conductivity of the working fluid
should also preferably be high in order to minimise the radial temperature
gradient and to reduce the possibility of nucleate boiling at the wick/wall
interface.

The resistance to fluid flow will be minimised by choosing fluids with low
values of vapour and liquid viscosity.

A convenient means for quickly comparing working fluids is provided by the Merit number, introduced in Chapter 2, defined as $\sigma_\ell \, L \, \rho_\ell / \mu_\ell$ where σ_ℓ is the surface tension, L the latent heat of vaporisation, ρ_ℓ the liquid density and μ_ℓ the liquid viscosity. Fig. 3.1 gives the Merit number M at the boiling point for working fluids covering temperature ranges beteeen 200 and 1750 K. One obvious feature is the marked superiority of water with its high latent heat and surface tension, compared with all the organic fluids, such as acetone and the alcohols. Final fluid selection is, of course also based on cost, availability, compatibility and the other factors listed above.

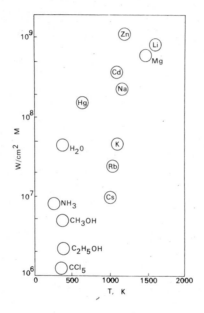

Fig. 3.1 Merit Number for selected working fluids at their boiling point (Courtesy Philips Technical Review)

As Asselman and Green (3.1) point out, a high Merit number is not the only criterion for the selection of the working fluid, and other factors may, in a particular situation, be of greater importance. For example on grounds of cost, potassium might be chosen rather than caesium or rubidium, which are one hundred times more expensive. Also, over the temperature range 1200 – 1800 K, lithium has a higher Merit number than most metals, including sodium. However its use requires a container made from an expensive lithium-resistant alloy, whereas sodium can be contained in stainless steel. It may therefore, be cheaper and more convenient to accept a lower performance heat pipe made from sodium/stainless steel.

Working fluids used in heat pipes range from helium at 4 K up to lithium at
2300 K. Fig. 3.1 shows the superiority of water over the range 350 - 500 K,
where the alternative organic fluids tend to have considerably lower Merit
numbers. At slightly lower temperatures, 270 - 350 K, ammonia is a desirable
fluid, although it requires careful handling to retain high purity, and
acetone and the alcohols are alternatives having lower vapour pressures.
These fluids are commonly used in heat pipes for space applications. Water
and methanol, both being compatible with copper, are often used for cooling
electronic equipment.

For temperatures over 500 K up to 650 K, the high temperature organic heat
transfer fluids such as Thermex (ICI) and Dowtherm A (Dow Chemical Co.) may
be used. Both of these fluids are diphenyl/diphenyl oxide eutectics having
boiling points around 260°C at atmospheric pressure. Unfortunately they have
a low surface tension and poor latent heat of vaporisation. As with many
other organic compounds, diphenyls are readily broken down when film temperatures exceed the critical value. However, unlike many other fluids having
similar operating temperature ranges, these eutectic mixtures have a specific
boiling point, rather than a boiling range.

Moving further up the temperature scale, one now enters the regime of liquid
metals. Mercury has a useful operating temperature range of about 500 - 950 K
and has attractive thermodynamic properties. It is also liquid at room
temperature, which facilitates handling, filling and start-up of the heat
pipe.[*]

Apart from its toxicity the main drawback to the use of mercury as a working
fluid in heat pipes, as opposed to thermal syphons, is the difficulty
encountered in wetting the wick and wall of the container. There are few
papers specifically devoted to this topic, but Deverall (3.2) at Los Alamos
and Reay (3.3) have both reported work on mercury wetting.
Bienert (3.4), in proposing mercury/stainless steel heat pipes for solar
energy concentrators, used Deverall's technique for wetting the wick in the

[*] There is one exception to this - sulphur has been proposed as a suitable
   fluid - see Section 3.5.

evaporator section of the heat pipe and achieved sufficient wetting for
gravity-assisted operation. He argued that non-wetting in the condenser
region of the heat pipe should enhance dropwise condensation which would
result in higher film coefficients than those obtainable with film condensation.

Moving yet higher up the vapour temperature range, caesium, potassium and
sodium are acceptable working fluids and their properties relevant to heat
pipes are well documented (see Appendix 1). Above 1400 K lithium is generally
a first choice as a working fluid, but silver has also been used (3.5).

3.2 The Wick or Capillary Structure

The selection of the wick for a heat pipe depends on many factors, several of
which are closely linked to the properties of the working fluid. Obviously
the prime purpose of the wick is to generate capillary pressure to transport
the working fluid from the condenser to the evaporator. It must also be able
to distribute the liquid around the evaporator section to any areas where
heat is likely to be received by the heat pipe. Often these two functions
require wicks of different form, particularly where the condensate has to be
returned over a distance of, say, one metre, in zero gravity.

It can be seen from Chapter 2 that the maximum capillary head generated by a
wick increases with decreasing pore size. The wick permeability, another
desirable feature, increases with increasing pore size, however. For homogeneous wicks there is an optimum pore size which is a compromise. There are
three main types in this context. Low performance wicks in horizontal and
gravity assisted heat pipes should permit maximum liquid flow rate by having a
comparatively large pore size, as with 100 or 150 mesh. Where pumping capability is required against gravity, small pores are needed. In space the
constraints on size and the general high power capability needed necessitates
the use of non-homogeneous or arterial wicks aided by small pore structures
for axial liquid flow.

Another feature of the wick which must be optimised is its thickness. The
heat transport capability of the heat pipe is raised by increasing the wick
thickness. However, the increased radial thermal resistance of the wick
created by this would work against increased capability, and would lower the
allowable maximum evaporator heat flux. The overall thermal resistance at

the evaporator also depends on the conductivity of the working fluid in the
wick. (Table 3.2 gives measured values of evaporator heat fluxes for various
wick/working fluid combinations.) Other necessary properties of the wick are
compatibility with the working fluid, and wettability. It should be easily
formed to mould into the wall shape of the heat pipe and should preferably be
of a form which enables repeatable performance to be obtained. It should be
cheap.

So-called homogeneous wicks take many forms; meshes, foams, felts, fibres
and sinters. Other types include grooves and arterial wicks, which may be
coupled with homogeneous wicks for circumferential liquid distribution.
Forms of wicks used in heat pipes are shown in Fig. 3.2.

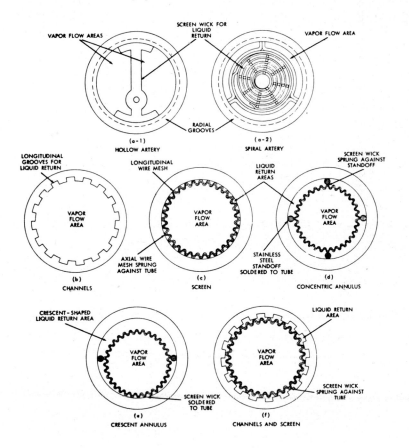

Fig. 3.2 Forms of wick used in heat pipes
(Courtesy NASA)

TABLE 3.2   MEASURED RADIAL EVAPORATOR HEAT FLUXES IN
HEAT PIPES

(These are not necessarily limiting values)

Working Fluid	Wick	Vapour Temp. ($^{\circ}$C)	Rad. Flux (W/cm^2)
Helium (3.6)	s/s mesh	-269	0.09
Nitrogen (3.6)	s/s mesh	-163	1.0
Ammonia (3.7)	various	20 - 40	5 - 15
Ethanol (3.8)	4 × 100 mesh s/s	90	1.1
Methanol (3.9)	nickel foam	25 - 30	0.03 - 0.4
" (3.9)	nickel felt	30	0.24 - 2.6
" (3.9)	1 × 200 mesh (horiz.)	25	0.09
" (3.9)	" (-2.5 cm head)	25	0.03
Water (3.7)	various	140 - 180	25 - 100
" (3.8)	mesh	90	6.3
" (3.8)	100 mesh s/s	90	4.5
" (3.9)	nickel felt	90	6.5
" (3.10)	sintered copper	60	8.2
Mercury (3.6)	s/s mesh	360	180
Potassium (3.6)	s/s mesh	750	180
" (3.7)	various	700 - 750	150 - 250
Sodium (3.6)	s/s mesh	760	230
" (3.7)	various	850 - 950	200 - 400
" (3.11)	3 × 65 mesh s/s	925	214
" (3.12)	508 × 3600 mesh s/s twill	775	1250
Lithium (3.6)	niobium 1% zirconium	1250	205
" (3.13)	niobium 1% zirconium	1500	115
" (3.13)	SGS - tantalum	1600	170
" (3.14)	W-26 Re grooves	1600	120
" (3.14)	W-26 Re grooves	1700	120
Silver (3.6)	tantalum 5% tungsten	-	410
" (3.14)	W-26 Re grooves	2000	155

3.2.1 Homogeneous structures. Of the wick forms available, meshes and twills are the most common. These are manufactured in a range of pore sizes and materials, the latter including stainless steel, nickel, copper and aluminium. Table 3.3 shows measured values of pore size and permeabilities for a variety of meshes and twills. Homogeneous wicks fabricated using metal foams, and more particularly felts, are becoming increasingly useful, and by varying the pressure on the felt during assembly, varying pore sizes can be produced. By incorporating removable metal mandrels, an arterial structure can also be moulded in the felt.

TABLE 3.3   WICK PORE SIZE AND PERMEABILITY DATA

Material and Mesh size	Capillary height [1] (cm)	Pore radius (cm)	Permeability (m^2)	Porosity (%)
Glass fibre (3.15)	25.4	–	0.061×10^{-11}	–
Refrasil sleeving (3.15)	22.0	–	0.104×10^{-10}	–
Refrasil (bulk) (3.16)	–	–	0.86×10^{-10}	–
Refrasil (batt) (3.16)	–	–	1.00×10^{-10}	–
Monel beads (3.17)				
30 - 40	14.6	0.052^2	4.15×10^{-10}	40
70 - 80	39.5	0.019^2	0.78×10^{-10}	40
100 - 140	64.6	0.013^2	0.33×10^{-10}	40
140 - 200	75.0	0.009	0.11×10^{-10}	40
Felt metal (3.18)				
FM1006	10.0	0.004	1.55×10^{-10}	–
FM1205	–	0.008	2.54×10^{-10}	–
Nickel powder (3.15)				
200μ	24.6	0.038	0.027×10^{-10}	–
500μ	>40.0	0.004	0.081×10^{-11}	–
Nickel fibre (3.15)				
0.01 mm dia.	>40.0	0.001	0.015×10^{-11}	68.9
Nickel felt (3.19)	–	0.017	6.0×10^{-10}	89
Nickel foam (3.19)				
Ampornik 220.5	–	0.023	3.8×10^{-9}	96
Copper foam (3.19)				
Amporcop 220.5	–	0.021	1.9×10^{-9}	91
Copper powder (sintered) (3.18)	156.8	0.0009	1.74×10^{-12}	52
" " " (3.20)				
45 - 56μ	–	0.0009	–	28.7
100 - 125μ	–	0.0021	–	30.5
150 - 200μ	–	0.0037	–	35
Nickel 50 (3.15)	4.8	–	–	62.5
50 (3.21)	–	0.0305	6.635×10^{-10}	–
Copper 60 (3.18)	3.0	–	8.4×10^{-10}	–
Nickel 60 (3.20)	–	0.009	–	–
100 (3.21)	–	0.0131	1.523×10^{-10}	–
100 (3.22)	–	–	2.48×10^{-10}	–
120 (3.18)	5.4	–	6.00×10^{-10}	–
120^3 (3.18)	7.9	0.019	3.50×10^{-10}	–
$2^5 \times 120$ (3.23)	–	–	1.35×10^{-10}	–
120 (3.24)	–	0.0102	–	–
S/s 180 $(22^\circ C)$ (3.25)	8.0	–	0.5×10^{-10}	–
2×180 $(22^\circ C)$ (3.25)	9.0	–	0.65×10^{-10}	–
200 (3.21)	–	0.0061	0.771×10^{-10}	–
200 (3.19)	–	–	0.520×10^{-10}	

.../Cont'd

Material and Mesh size	Capillary height (cm)	Pore radius (cm)	Permeability (m^2)	Porosity (%)
Nickel 200 (3.15)	23.4	0.004	0.62×10^{-10}	68.9
2 × 200 (3.23)	-	-	0.81×10^{-10}	-
Phosp./bronze 200 (3.26)	-	0.003	0.46×10^{-10}	67
Titanium 2 × 200 (3.20)	-	0.0015	×	67
4 × 200 (3.20)	-	0.0015	×	68.4
250 (3.22)	-	-	0.302×10^{-10}	-
Nickel[3] 2 × 250 (3.20)	-	0.002	×	66.4
4 × 250 (3.20)	-	0.002	×	66.5
325 (3.24)	-	0.0032	-	-
Phosp./bronze (3.26)	-	0.0021	0.296×10^{-10}	67
S/s (twill) 80[4](3.27)	-	0.013	2.57×10^{-10}	-
90[4](3.27)	-	0.011	1.28×10^{-10}	-
120[4](3.27)	-	0.008	0.79×10^{-10}	-
250 (3.24)	-	0.0051	-	-
270 (3.24)	-	0.0041	-	-
400 (3.24)	-	0.0029	-	-
450 (3.27)	-	0.0029	-	-

[1] Obtained with water unless stated otherwise
[2] Particle diameter
[3] Oxidised
[4] Permeability measured in direction of warp
[5] Denotes number of layers

Fibrous materials have been widely used in heat pipes, and generally have
small pore sizes. The main disadvantage is that ceramic fibres have little
stiffness and usually require a continuous support for example by a metal
mesh. Thus while the fibre itself may be chemically compatible with the
working fluids, the supporting materials may cause problems.

Sintered powders are available in spherical form in a number of materials,
and fine pore structures may be made, possibly incorporating larger arteries
for added liquid flow capability. Leaching has been used to produce fine
longitudinal channels, and grooved walls in copper and aluminium heat pipes
have been applied for heat pipes in zero-gravity environments. (In general
grooves are unable to support significant capillary heads alone in earth
gravity applications. Also entrainment may limit the axial heat flow.)

(Covering grooves with a mesh prevents this).

3.2.2 **Arterial wicks.**   Arterial wicks are necessary in high performance
heat pipes for spacecraft, where temperature gradients into the heat pipe
have to be minimised to counter the adverse effect of what are generally low
thermal conductivity working fluids. An arterial wick developed at IRD is
shown in Fig. 3.3. The bore of the heat pipe in this case was only 5.25 mm.
This heat pipe, developed for the European Space Organisation (ESRO) was
designed to transport 15 W over a distance of 1 m with an overall temperature
drop not exceeding 6°C. The wall material was aluminium alloy and the
working fluid acetone. (See design example A at end of this chapter).

The aim of this wick system was to obtain liquid transport along the pipe
with the minimum pressure drop. A high driving force was achieved by
covering the six arteries with a fine screen.

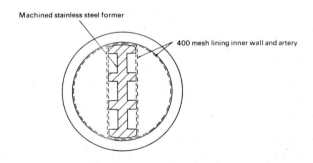

Machined stainless steel former

400 mesh lining inner wall and artery

Fig. 3.3 Arterial wick developed at IRD (Courtesy IRD)

In order to achieve the full heat transport potential of the arterial wick,
the artery must be completely shut off from the vapour space. The maximum
capillary driving force is thus determined by the pore size of the screen.
Thus a high degree of quality control was required during the manufacturing
process to ensure that the artery was successfully closed and the screen
undamaged.

A further consideration in the design of arterial heat pipes is that of
vapour or gas blockage of the arteries. If a vapour or gas bubble forms
within or is vented into the artery, then the transport capability is
seriously reduced. Indeed, if the bubble completely blocks the artery then

the heat transport capability is dependent on the effective capillary radius
of the artery, i.e. there is an effective state of open artery. In order
that the artery will reprime following this condition the heat load must be
reduced to a value below the maximum capability associated with the open
artery. (See also Chapter 6).

The implications of wick design and working fluid properties in arterial heat
pipes are as follows:

(i)   The working fluid must be thoroughly degassed prior to
        filling to minimise the risk of non-condensable gases
        blocking the artery.

(ii)   The artery must not be in contact with the wall to prevent
        nucleation within it.

(iii)   A number of redundant arteries should be provided to allow
        for some degree of failure.

(iv)   Successful priming (i.e. refilling) of the artery, if
        applied to spacecraft, must be demonstrated in a one 'g'
        environment, it being expected that priming in a zero 'g'
        environment will be easier.

3.3 Thermal Resistance of Saturated Wicks

One feature mentioned in discussions of the desirable properties of both the
wick and the working fluid is the thermal conductivity. Expressions are
available for predicting the thermal conductivities of saturated wicks of
several types, and these are discussed below. The conductivity is an important factor in determining the allowable wick thickness.

3.3.1 Meshes. Gorring and Churchill (3.28) present solutions for the
determination of the thermal conductivity of heterogeneous materials which
are divided into three categories; dispersions, packed beds, and continuous
pairs. No satisfactory solution for a mesh is given because a mesh is a
limiting case of dispersion, i.e. the particles are in contact but not tightly
packed. However since the conductivity of dispersions is less than that of
packed beds, an estimate of mesh conductivity can be made using Rayleigh's
expression for the effective conductivity of a dispersion consisting of a

square array of uniform cylinders:

$$k_w = (\frac{\beta - \varepsilon}{\beta + \varepsilon}) k_\ell \qquad\qquad\qquad \ldots 3.1$$

where $\beta = (1 + \frac{k_s}{k_\ell}) / (1 - \frac{k_s}{k_\ell})$

k_s = thermal conductivity of solid phase

k_ℓ = thermal conductivity of liquid phase

ε = volume fraction of solid phase

An example of the use of this equation is given in Design Calculation A. at the end of this Chapter.

3.3.2 Sintered wicks.   The exact geometric configuration of a sintered wick is unknown because of the random dispersion of the particles and the varying degree of deformation and fusion which occurs during the sintering process (see Chapter 4).   For this reason it is suggested that the sintered wick be represented by a continuous solid phase containing a random dispersion of randomly sized spheres of liquid.

Maxwell (3.29) has derived an expression which gives the thermal conductivity of such a heterogeneous material:

$$k_w = k_s \left[\frac{2 + k_\ell/k_s - 2\varepsilon (1 - k_\ell/k_s)}{2 + k_\ell/k_s + \varepsilon (1 - k_\ell/k_s)} \right] \qquad \ldots 3.2$$

Gorring and Churchill show that this expression agrees reasonably well with experimental results.

3.3.3 Grooved wicks. The radial thermal resistance of grooves will be radically different in the evaporator and condenser sections. This occurs because of the differences in the mechanisms of heat transfer.   In the evaporator the land or fin tip plays no active part in the heat transfer process. The probable heat flow path is conduction via the fin, conduction across a liquid film at the meniscus and evaporation at the liquid vapour interface.

In the condenser section goooves will be flooded and the fin tip plays an
active role in the heat transfer process. The build up of a liquid film at
the fin tip will provide the major resistance to heat flow.

The thickness of the liquid film is a function of the condensation rate and
the wetting characteristics of the working fluid.

Since the mechanism in the evaporator section is less complex and should
provide the greatest resistance we will concentrate on the analysis of that
region.

Joy (3.30), and Eggers and Serkiz (3.31) propose identical models which assume
one dimensional heat conduction along the fin and one dimensional conduction
near the fin tip across the liquid to the liquid/vapour interface where
evaporation occurs. In the liquid the average heat flow length is taken to
be a quarter of the channel width and the heat flow area the channel half
width times the input length.

Thus

$$\frac{\Delta T}{Q} = \frac{a}{k_s \, N f \, \ell_e} + \frac{1}{4 \, k_\ell \, \ell_e \, N} + \frac{1}{h_e \, \pi \, b \, N \, \ell_e} \quad \text{... 3.3}$$

where   N   is the number of channels

        a   is the channel depth

        b   is the channel half width.

        f   is the fin thickness

Kosowski and Kosson (3.32) have made measurements of the maximum heat transport capability and radial thermal resistance of an aluminium grooved heat
pipe using Freon 21, Freon 113 and ammonia as the working fluids. The
relevant dimensions of their heat pipes were as follows:

$$N = 30$$
$$a = 0.89 \text{ mm}$$
$$2b = 0.76 \text{ mm}$$
$$\text{Pipe outside dia.} = 12.7 \text{ mm}$$

Heat pipe No. 1 ℓ_e = 304.8 mm, ℓ_c = 477.6 mm

Heat pipe No. 2 ℓ_e = 317.5 mm, ℓ_c = 503 mm

The following heat transfer coefficients (based upon the outside area) were measured:

Fluid	h_e W/m^2 $^\circ C$	h_c W/m^2 $^\circ C$
Freon 21 (heat pipe No. 1)	1134	1700
Freon 113 (heat pipe No. 2)	652	1134
Ammonia (heat pipe No. 3)	2268	2840

Converting the heat transfer coefficients into a thermal resistance:

Fluid	R $^\circ$C/W (evap)	R $^\circ$C/W (cond)
Freon 21	0.0735	0.031
Freon 113	0.122	0.044
Ammonia	0.035	0.0175

The contribution due to fin conduction is 0.0018°C/W, (f = 0.25 mm, k_w = 220 W/m$^\circ$C) and is negligible. This bears out Kosowski and Kosson's observation that the percentage fill has little effect upon the thermal resistance.

The evaporation term is also small, and the most significant contribution to the resistance is the liquid conduction term. Comparing the theory and experiment, results suggest that the theory over-predicts the conduction resistance by a considerable amount (50 - 300%). It would therefore be more accurate to use the integrated mean heat flow length $(1 - \frac{\pi}{4})b$, rather than $\frac{b}{2}$ such that:

$$\frac{\Delta T}{Q} = \frac{(1 - \frac{\pi}{4})}{2 k_\ell \ell_e N} \qquad \qquad \ldots 3.4$$

knowing the duty and allowable temperature drop into the vapour space, the
number of grooves can be calculated for various geometries and working fluids.

In the heat pipe condenser section, or when the channels are mesh-covered, the
fin tip plays an active part in the heat transfer process, and the channels
are completely filled. In this case the parallel conduction equation is used:

$$k_w = k_s \{1 - \varepsilon (1 - \frac{k_\ell}{k_s})\} \qquad \ldots \ 3.5$$

where ε, the liquid void fraction, is given by:

$$\varepsilon = \frac{2b}{2b + f} \quad \text{for channels}$$

where f is the fin thickness.

3.3.4 Concentric annulus.

In this case the capillary action is derived from
a thin annulus containing the working fluid. Thus:

$$k_w = k_\ell \qquad \ldots \ 3.6$$

This case may also be used to analyse the effects of loose fitting mesh and
sintered wicks.

3.4 The Container

The function of the container is to isolate the working fluid from the outside environment. It has therefore to be leak-proof, maintain the pressure
differential across its walls, and enable the transfer of heat to take place
into and from the working fluid.

Selection of the container material depends on several factors. These are as
follows:

(i)   Compatibility (both with working fluid and the external
       environment).

(ii)   Strength to weight ratio.

(iii) Thermal conductivity.

(iv) Ease of fabrication, including weldability, machineability, ductility.

(v) Porosity.

(vi) Wettability.

Most of these are self-explanatory. A high strength to weight ratio is more important in spacecraft applications, and the material should be non-porous to prevent the diffusion of gas into the heat pipe. A high thermal conductivity ensures minimum temperature drop between the heat source and the wick.

The thermal conductivity of some wall materials is given in Appendix 2.

3.5 Compatibility

Compatibility has already been discussed in relation to the selection of the working fluid, wick and containment vessel of the heat pipe. However, this feature is of prime importance and warrents particular attention here.

The two major results of incompatibility are corrosion and the generation of non-condensable gas. If the wall or wick material is soluble in the working fluid, mass transfer is likely to occur between the condenser and evaporator, with solid material being deposited in the latter. This will result either in local hot spots or blocking of the pores of the wick. Non-condensable gas generation is probably the most common indication of heat pipe failure and, as the non-condensables tend to accumulate in the heat pipe condenser section, which gradually becomes blocked, it is easy to identify because of the sharp temperature drop which exists at the gas/vapour interface.

Some compatibility data is of course available in the general scientific publications and from trade literature on chemicals and materials. However, it has become common practice to carry out life tests on representative heat pipes, the main aim of these tests being to estimate long term materials compatibility under heat pipe operating conditions. At the termination of life tests gas analyses and metallurgical examinations, as well as chemical analysis of the working fluid may be carried out. (See also Section 4.2, Chapter 4).

Many laboratories have carried out life tests, and a vast quantity of data
has been published. However it is important to remember that while life test
data obtained by one laboratory may indicate satisfactory compatibility,
different assembly procedures at another laboratory, involving for example
a non standard materials treatment process, may result in a different
corrosion or gas generation characteristic. Thus it is important to obtain
compatibility data whenever procedural changes in cleaning or pipe assembly
are made.

Stainless steel is a suitable container and wick material for use with
working fluids such as acetone, ammonia, and liquid metals from the point of
view of compatibility. Its low thermal conductivity is a disadvantage, and
copper and aluminium are used where this feature is important. The former is
particularly attractive for mass-produced units using water as the working
fluid. Plastic has been used as the container material, and at very high
temperatures ceramics and refractory metals such as tantalum have been given
serious consideration. In order to introduce a degree of flexibility in the
heat pipe wall, stainless steel bellows have been used, and in cases where
electrical insulation is important, a ceramic or glass-to-metal seal has been
incorporated. This must of course be used in conjunction with electrically
non-conducting wicks and working fluids.

A comprehensive review of material combinations in the intermediate temperature range has been carried out by Basiulis and Filler (3.33) and is
summarised below. Results are given in the paper over a wider range of
organic fluids, most produced by Dow Chemicals, than given in Table 3.4.

Tests in excess of 8000 hours with ammonia/alumium were reported but only
1008 hours had been achieved at the same time of data compilation with
aluminium/acetone. No temperatures were specified by Basiulis for these
tests: other workers have now exceeded 16 000 hours with the latter
combination.

More recently the life test work at IKE, Stuttgart has been published (3.34),
involving tests on about 40 heat pipes. The tests indicated that copper/
water heat pipes could be operated without degradation over long periods (now
exceeded 20 000 hr), but severe gas generation was observed with stainless

TABLE 3.4   COMPATIBILITY DATA
(Low temperature working fluids)

Wick Material	Working Fluids					
	Water	Acetone	Ammonia	Methanol	Dow-A	Dow-E
Copper	RU	RU	NR	RU	RU	RU
Aluminium	GNC	RL	RU	NR	UK	NR
Stainless Steel	GNT	PC	RU	GNT	RU	RU
Nickel	PC	PC	RU	RL	RU	RL
Refrasil Fibre	RU	RU	RU	RU	RU	RU

RU   Recommended by past successful usage

RL   Recommended by literature

PC   Probably compatible

NR   Not recommended

UK   Unknown

GNC   Generation of gas at all temperatures

GNT   Generation of gas at elevated temperatures, when
      oxide present.

steel/water heat pipes. IKE had some reservations concerning acetone with
copper and stainless steel. While compatible, it was stressed that proper
care had to be given to the purity of both the acetone and the metal. The
same reservation applied to the use of methanol.

Exhaustive tests on stainless steel/water heat pipes were also carried out at
Ispra (3.35), where experiments were conducted with vapour temperatures as
high as 250°C.

It was found that neither variation of the fabrication parameters nor the
addition of a large percentage of oxygen to the gas plug resulted in a
drastic reduction of the hydrogen generation at 250°C. Hydrogen was generated
within 2 hours of start-up in some cases. The stainless steel used was type
316 and such procedures as passivation and out-gassing were ineffective in
arresting generation. However, it was found that the complimentary formation
of an oxide layer on the steel did inhibit further hydrogen generation.

Gerrels and Larson (3.36), as part of a study of heat pipes for satellites, also carried out comprehensive life tests to determine the compatibility of a wide range of fluids with aluminium (6061 alloy) and stainless steel (type 321). The fluids used included ammonia, which was found to be acceptable. It is important, however, to ensure that the water content of the ammonia is very low, only a few ppm concentration being acceptable with aluminium and stainless steel.

The main conclusions concerning compatibility made by Gerrels and Larson are given below, data being obtained for the following fluids:

n-pentane	CP-32 (Monsanto experimental fluid)
n-heptane	CP-34 " " "
benzene	ethyl alcohol
toluene	methyl alcohol
water (with stainless steel)	ammonia
Freon 11	n-butane
Freon 113	

The stainless steel used with the water was type 321.

All life tests were carried out in gravity refluxing containers, with heat being removed by forced air convection, and being put in by immersion of the evaporator sections in a temperature-controlled oil bath.

Preparation of the aluminium alloy was as follows: Initial soak in a hot alkaline cleaner followed by deoxidation in a solution of 112 gm sodium sulphate and 150 ml concentrated nitric acid in 850 ml water for 20 min. at 60°C. In addition the aluminium was either machined or abraded in the area of the welds. A mesh wick of commercially pure aluminium was inserted in the heat pipes. The capsules were TIG welded under helium in a vacuum purged inert gas welding chamber. Leak detection followed welding, and the capsules were also pressure tested to 70 bar. A leak check also followed the pressure test.

The type 321 stainless steel container was cleaned before fabrication by soaking in hot alkaline cleaner and pickling for 15 min. at 58°C in a

solution of 15% by volume of concentrated nitric acid, 5% by volume of
concentrated hydrochloric acid, and 80% water. In addition the stainless
steel was passivated by soaking for 15 min. at $65^{\circ}C$ in a 15% solution of
nitric acid. Type 316 stainless steel was used as the wick. The capsule was
TIG welded in air with argon purging*.

The boiling off technique was used to purge the test capsules of air.

In the case of methyl alcohol, reaction was noted during the filling
procedure, and a full life test was obviously not worth proceeding with.

Sealing of the capsules was by a pinch-off, followed by immersion of the
pinched end in an epoxy resin for final protection.

The following results were obtained:

n-pentane: Tested for 750 hours at $150^{\circ}C$. Short term instabilities
            noted with random fluctuation in temperature of $0.2^{\circ}C$.
            On examination of the capsule very light brownish areas
            of discolouration were noted on the interior wall, but
            the screen wick appeared clean. No corrosion evidence
            was found. The liquid removed from the capsule was found
            to be slightly brown in colour.

n-heptane Tested for 600 hours at $160^{\circ}C$. Slight interior resistances
            noted after 465 hours, but on opening the capsule at the
            end of the tests, the interior, including the screen, was
            clear, and the working fluid clean.

Benzene: Tested for 750 hours at $150^{\circ}C$ plus. Vapour pressure
            6.7 bar. Very slight local areas of discolouration found
            on the wall, the wick was clean, there was no evidence of
            corrosion and the liquid was clear. It was concluded that
            benzene was very stable with the chosen aluminium alloy.

* In both series of U.S.A. tests mixtures of materials were used. This is
not good practice in life tests as any degradation may not be identified as
being caused by the particular single material.

HP—E

Toluene: Test run for 600 hours at 160°C. A gradual decrease in
            condenser section temperature was noted over the first
            200 hours of testing, but no change was noted following
            this. On opening the capsule, slight discolouration was
            noted locally on the container wall. This seemed to be a
            surface deposit, with no signs of attack on the aluminium.
            The screen material was clean and the working fluid clear
            at the end of the test.

Water (stainless steel): Tested at 150°C plus for 750 hours. Vapour
            pressure 6.7 bar. Large concentration of hydrogen was
            found when analysis of the test pipe was performed. This
            was attributed in part to poor purging procedure as there
            was discoloration in the area of the welds, and the authors
            suggest that oxidation of the surfaces had occurred. A
            brown precipitate was also found in the test heat pipe.

Freon 11:   Two capsules tested, one for 500 hours at 68°C, the second
            for 500 hours at 95°C. When the first capsule was opened
            a few small areas of discolouration were observed on the
            inner wall. The screen appeared clean and the fluid was
            clear. The second Freon 11 capsule was clean throughout
            the interior following the test and the fluid was clear.

Freon 113: Two capsules tested at the same temperatures over identical
            periods to the Freon 11 examples. The interiors of both
            capsules were clean following these tests and the fluid
            was clear.

CP-32   Tested for 550 hours at 158°C. A brownish deposit was
            found locally on interior surface. Screen clean, but the
            working fluid was darkened.

CP-34:   Tested for 550 hours at 158°C. Gas generation was noted.
            Also extensive local discolouration on the capsule wall
            near the liquid surface. No discolouration on the screen.
            The fluid was considerably darkened.

Ammonia: Tested at 70°C for 500 hours. Some discolouration of the
           wall and mesh was found following the tests. This was
           attributed to some non-volatile impurity in the ammonia
           which could have been introduced when the capsule was
           filled. In particular the lubricant on the valve at the
           filling position could have entered with the working fluid.
           (This was the only test pipe on which the filling was
           performed through a valve).

n-Butane: Tested for 500 hours at 68°C. It was considered that there
           could have been non-condensable gas generation, but the
           fall off in performance was attributed to some initial
           impurity in the n-butane prior to filling. The authors
           felt that this impurity could be isobutane. Further tests
           on purer n-butane gave better results, but the impurity
           was not completely removed.

Gerrels and Larsen argued thus concerning the viability of their life tests:
"It should be emphasised that the present tests were planned to investigate
the compatibility of a particular working fluid-material combination for long
term (5 year) use in vapour chamber radiator under specified conditions. The
reference conditions call for a steady state temperature of 143°C for the
primary radiator fluid at the inlet to the radiator and a 160°C short term
peak temperature. The actual temperature to which the vapour chamber working
fluid is exposed must be somewhat less than the primary radiator fluid temperature, since some temperature drop occurs from the primary radiator fluid
to the evaporative surface within the vapour chamber. It is estimated that
in these capsule tests the high temperature fluids were exposed to temperatures at least 10°C higher than the peak temperature and at least 20°C
greater than long-term steady state temperatures that the fluids would experience in the actual radiator. Although the time of operation of these
capsule tests is only about 1% of the planned radiator lifetime, the conditions of exposure were much more severe. It seems reasonable then, to assume
that if the fluid-material combination completed the capsule tests with no
adverse effects, it is a likely candidate for a radiator with a 5 year
lifetime."

On the basis of the above tests, Gerrels and Larsen selected the following
working fluids:

6061 Aluminium at temperatures not exceeding 150°C:

       Benzene

       n-Heptane

       n-Pentane

6061 Aluminium at temperatures not exceeding 94°C:

       Freon 11

       Freon 113

6061 Aluminium at temperatures not exceeding 65°C:

       Ammonia

       n-Butane

Data from Dupont indicates that the corrosion rate of Freon 11 with aluminium
is 1.25 x 10^{-6} cm per month at 115°C. Other tests on Freon 113 show no
corrosion (3.37) with aluminium in 100 hours at the boiling point. Freon 113
has been stored for 2 years with various metals at 150°C with 0.3 to 0.4%
decomposition.

The following fluids were felt not to be suitable:

       Water (in type 321 stainless steel)

       CP-32

       CP-34

       Methyl alcohol   in 6061 aluminium

       Toluene

Gerrels and Larsen point out that Los Alamos laboratory obtained heat pipe
lives in excess of 3000 hours without degradation using a combination of
water and type 347 stainless steel.

Other data (3.38) suggests that alcohols in general are not suitable with
aluminium.

Summarising the data of Gerrels and Larsen, ammonia was recommended as the
best working fluid for vapour chambers operating below 65°C and n-pentane the

best for operation above this temperature, assuming that aluminium is the
container material.

At the other end of the temperature scale, long lives have been reported (3.5)
for heat pipes with lithium or silver as the working fluid. With a tungsten-
rhenium (W - 26 Re) container, a life of many years was forecast with lithium
as the working fluid, operating at 1600°C. At 1700°C significant corrosion
was observed after one year, while at 1800°C the life was as short as one
month. W-26 Re/silver heat pipes were considered capable of operating at
2000°C for 1000 hr. Some other results are presented in Table 3.5.

TABLE 3.5 COMPATIBILITY DATA
(Life tests on high temperature heat pipes)

Working Fluid	Material		Vapour Temp.(°C)	Duration (h)
	Wall	Wick		
Caesium	Ti		400	> 2 000
	Nb + 1% Zr		1100	184
Potassium	Ni		600	> 6 000
	Ni		600	16 000
	Ni		600	>24 500
Sodium	Hastelloy X		715	> 8 000
	Hastelloy X		715	>20 000
	316 SS		771	> 4 000
	Nb + 1% Zr		850	>10 000
	Nb + 1% Zr		1100	1 000
Bismuth	Ta		1600	39
	W		1600	118
Lithium	Nb + 1% Zr		1100	4 300
	Nb + 1% Zr		1500	> 1 000
	Nb + 1% Zr		1600	132
	Ta		1600	17
	W		1600	1 000
	SGS - Ta		1600	1 000
	TMZ		1500	9 000
Lead	Nb + 1% Zr		1600	19
	SGS - Ta		1600	1 000
	W		1600	1 000
	Ta		1600	> 280
Silver	Ta		1900	100
	W		1900	335
	W		1900	1 000
	Re	W	2000	300

One subject of much argument is the method of conducting life tests and their validity when extrapolating likely performance over a period of several years. For example, on satellites, where remedial action in the event of failure is difficult, if not impossible, to implement, a life of 7 years is a standard minimum requirement.* It is therefore necessary to accelerate the life tests so that reliability over a longer period can be predicted with a high degree of accuracy.

Life tests on heat pipes are commonly regarded as being primarily concerned with the identification of any incompatibilities which may occur between the working fluid and wick and wall materials. However, the ultimate life test would be in the form of a long term performance test under likely operating conditions. However, if this is carried out, it is difficult to accelerate the life test by increasing, say, the evaporator heat flux, as any signif- icant increase is likely to cause dry-out as the pipe will be operating well in excess of its probable design capabilities. Therefore any accelerated life test which involves heat flux increases of the order of, say, four over that required under normal operating conditions must be carried out in the reflux mode, with regular performance tests to ensure that the design capa- bility is still being obtained.

An alternative possibility as a way of accelerating any degradation processes, and one which may be carried out with the evaporator up if the design permits, is to raise the operating temperature of the heat pipe. One draw- back of this method is the effect increased temperature may have on the stability of the working fluid itself. Acetone cracking, for example, might be a factor where oxides of metals are present, resulting in the formation of diacetone alcohol, which has a much higher boiling point than pure acetone.

Obviously there are many factors to be taken into account when preparing a life test programme, including such questions as the desirability of heat pipes with valves, or completely sealed units as used in practice. This topic is of major importance, and life test procedures are discussed more fully in Chapter 4.

* European Space Research Organisation requirement

A considerable amount of new compatibility data was presented at the Bologna
Heat Pipe Conference. Of interest also was the work by Polasek and Stulc on
the use of sulphur as a working fluid, and this is also described below. (3.48)

One of the most comprehensive life test programmes is that being carried out
by Hughes Aircraft Co. (3.45). Additional experience has been gained with some
of the material combinations discussed above, some tests now having lasted
nine years, and a summary of recommendations based on these tests is given in
Table 3.6.

TABLE 3.6. HUGHES AIRCRAFT COMPATIBILITY RECOMMENDATIONS

	Recommended	Not Recommended
Ammonia	Aluminium Carbon Steel Nickel Stainless Steel	Copper
Acetone	Copper Silica Aluminium * Stainless Steel *	
Methanol	Copper Stainless Steel Silica	Aluminium
Water	Copper Monel 347 Stainless Steel **	Stainless Steel Aluminium Silica Inconel Nickel Carbon Steel
Dowtherm A (Thermex)	Copper Silica Stainless Steel ***	
Potassium	Stainless Steel Inconel	Titanium
Sodium	Stainless Steel Inconel	Titanium

Note: Type 347 stainless steel as specified in AISI codes does not contain
tantalum. AISI Type 348, which is otherwise identical except for a small
tantalum content, should be used in the UK. (Authors). This also applies to
the comments on Page 126.

* The use of acetone with aluminium and/or stainless steel presented problems
  to the authors, but others have had good results with these materials. The
  problem may be temperature-related; use with caution.

** Recommended with reservations. Investigation of the suitability of the
   water/347 stainless steel combination is continuing.

*** This combination should be used only where some non-condensible gas in the
    heat pipe is tolerable, particularly at higher temperatures

The lack of support given to a nickel/water combination is based on an Arrhenius type accelerated life test carried out by Anderson (3.46). Work
carried out on nickel wicks in water heat pipes, generally with a copper wall,
has, in the authors' experience, not created compatibility problems, and this
area warrants further investigation. With regard to water/stainless steel combinations, for some years the subject of considerable study and controversy,
the Hughes work suggests that type 347 stainless steel is acceptable as a container with water. Tests have been progressing since December 1973 on a 347
stainless steel container, copper wicked water heat pipe operating at 165°C,
with no trace of gas generation. (Type 347 stainless steel contains no titanium,
but does contain tantalum, which is a recognised hydrogen getter). Surprisingly,
a type 347 wick caused rapid gas generation. The use of Dowtherm A (equivalent
to Thermex, manufactured by ICI) is recommended for moderate temperatures only,
breakdown of the fluid progressively occurring above about 160°C. With careful
materials preparation Thermex appears compatible with mild steel and the Hughes
data is limited somewhat by the low operating temperature conditions.

Hughes emphasized the need to carry out rigorous and correct cleaning procedures, and also stressed that the removal of cleaning agents and solvents prior
to filling with working fluid is equally important.

Geiger and Quataert (3.47) have carried out corrosion tests on heat pipes
using tungsten as the wall material and silver (Ag), gold (Au), copper (Cu),
gallium (Ga), germanium (Ge), indium (In) and tin (Sn) as working fluids. The
results from tests carried out at temperatures of up to 2650°C enable Table
3.5 to be extended above 2000°C, albeit for heat pipes having comparatively
short lives. Of the above combinations, tungsten and silver proved the most
satisfactory, giving a life of 25 hours at 2400°C, with a possible extension
if improved quality tungsten could be used.

Sulphur - A New Working Fluid: Polasek and Stulc have published data (3.48), describing the use of sulphur as a heat pipe working fluid for the operating temperature range 200°C to 600°C (the most difficult to satisfy). Pure sulphur has a melting point of 112°C, a boiling point of 444°C and a critical tempera- ture of 1040°C. However, some of its properties, particularly its viscosity, are extremely temperature sensitive, making it unsatisfactory as a heat pipe working fluid for use over a wide temperature range, as shown in Fig. 3.4.

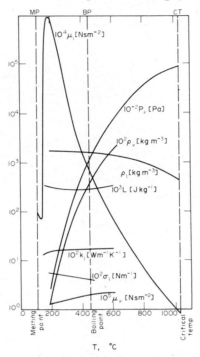

Fig. 3.4   Effect of temperature on the properties of sulphur

Polasek found that the addition of a small quantity of iodine to the sulphur significantly changed the dynamic viscosity of sulphur. Without the iodine, Polasek observed that the sulphur was too viscous to flow at a temperature of 187°C, but if heated to 300°C, the viscosity reduced rapidly to values of the same order as those measured at about 120°C. The effect on the viscosity of adding iodine in proportions of 3, 5 and 10 per cent by weight is shown in Fig. 3.5. It can be seen that there is a dramatic reduction in viscosity, and the peak described above is completely eliminated, as is the discontinuity associated with the permitted heat pipe operating temperature range.

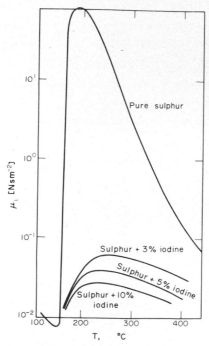

Fig. 3.5 Effect of added iodine on the dynamic viscosity
          of sulphur

Polasek and Stulc constructed a number of stainless steel/sulphur gravity-
assisted heat pipes having an outside diameter of 3.2 cm and a length of 2 m,
of which 0.5 m was used as the evaporator section and 1 m the condenser. Axial
heat fluxes of the order of 600 W/cm^2 were obtained at the top end of the
operating temperature range , of the same order as that obtained with Dowtherm
A (Thermex), although of course the organic fluid cannot be used at such temperatures for any extended period due to degradation.

This work is certain to encourage further experiments using sulphur heat pipes,
including operation against gravity, quantification of the effect of iodine on
the other properties of sulphur, and life tests.

3.6 Fluid Inventory

A feature of heat pipe design which is important when considering small heat pipes and units for space use is the working fluid inventory. It is common practice to include a slight excess of working fluid over and above that required to saturate the wick, but when the vapour space is of small volume a noticeable temperature gradient can exist at the condenser, similar to that indicating the presence of non-condensable gas. This reduces the effective length of the condenser, hence impairing heat pipe performance.

Another drawback of excess fluid is peculiar to heat pipes in space, where in zero-g the fluid can move about the vapour space, affecting the dynamics of the spacecraft.

If there is a deficiency of working fluid the heat pipe may fail because of the inability of an artery to fill. This is not so critical with homogeneous wicks as some of the pores will still be able to generate capillary action. Marcus (3.39) discusses in detail these effects and the difficulties encountered in ensuring that the correct amount of working fluid is injected into the heat pipe.

One way of overcoming the problem is to provide an excess fluid reservoir, which behaves as a sponge, absorbing working fluid which is not required by the primary wick structure.

3.7 Priming

With heat pipes having some form of arterial wick, it is necessary to ensure that should an artery become depleted of working fluid, it should be able to refill automatically.

It is possible to calculate the maximum diameter of an artery to ensure that it will be able to re-prime:

The maximum priming height which can be achieved by a capillary is given by the equation:

$$h + h_c = \frac{\sigma_\ell \cos \theta}{(\rho_\ell - \rho_v)g} \times (\frac{1}{r_{pl}} + \frac{1}{r_{p2}}) \qquad \ldots 3.2$$

where h is the vertical height to the base of the artery

h_c is the vertical height to the top of the artery

r_{pl} is the first principal radius of curvature of the priming meniscus

r_{p2} is the second principal radius of curvature of the priming meniscus.

For the purpose of priming the second principal radius of curvature of the meniscus is extremely large (approximately $\ell \sin \phi$). For a cylindrical artery: $h_c = d_a$

and $r_{pl} = \dfrac{d_a}{2}$

where d_a is the artery diameter.

Hence the above equation becomes:

$$h + d_a = \frac{2\sigma_\ell \cos \theta}{(\rho_\ell - \rho_v)g \times d_a} \qquad \ldots 3.8$$

which produces a quadratic in d_a which may be solved as:

$$d_a = \frac{1}{2} \left[(\sqrt{h^2 + \frac{8\sigma_\ell \cos \theta}{(\rho_\ell - \rho_v)g}}) - h \right] \qquad \ldots 3.9$$

3.8 Heat Pipe Start-up Procedure

Heat pipe start-up behaviour is difficult to predict and may vary considerably depending upon many factors. The effects of working fluid and wick behaviour and configuration on start-up performance have been studied qualitatively, and a general description of start-up procedure has been obtained (3.40).

During start-up, vapour must flow at a relatively high velocity to transfer
heat from the evaporator to the condenser, and the pressure drop through the
centre channel will be large. Since the axial temperature gradient in a heat
pipe is determined by the vapour pressure drop, the temperature of the
evaporator will be initially much higher than that of the condenser. The
temperature level reached by the evaporator will, of course, depend on the
working fluid used. If the heat input is large enough, a temperature front
will gradually move towards the condenser section. During normal heat pipe
start-up, the temperature of the evaporator will increase by a few degrees
until the front reaches the end of the condenser. At this point the condenser
temperature will increase until the pipe structure becomes almost isothermal
(when lithium or sodium are used as working fluids, this process occurs at
temperature levels where the heat pipe becomes red hot, and the near isothermal behaviour is visible).

Heat pipes with screen covered channels behave normally during start-up as
long as heat is not added too quickly. Kemme found that heat pipes with
open channels did not exhibit straightforward start-up behaviour. Very large
temperature gradients were measured, and the isothermal state was reached in
a peculiar manner. When heat was first added, the evaporator temperature
levelled out at 525°C (sodium being the working fluid) and the front, with a
temperature of 490°C, extended only a short distance into the condenser
section. In order to achieve a near isothermal condition more heat was added,
but the temperature of the evaporator did not increase uniformly, a temperature of 800°C being reached at the end of the evaporator farthest from the
condenser. Most of the evaporator remained at 525°C and a sharp gradient
existed between these two temperature regions.

Enough heat was added so that the 490°C front eventually reached the end of
the condenser. Before this occurred, however, temperatures in excess of 800°C
were observed over a considerable portion of the evaporator. Once the condenser became almost isothermal, its temperature rapidly increased and the very
hot evaporator region quickly cooled in a pattern which suggested that liquid
return flow was in fact taking place. From this point the heat pipe behaved
normally.

In some instances during start-up, when the vapour density is low and its

velocity high, the liquid can be prevented from returning to the evaporator.
This is more likely to occur when open return channels are used for liquid
transfer than when porous media are used.

More recent work by van Andel (3.41) on heat pipe start-up has enabled some
quantitative relationships to be obtained which assist in ensuring that
satisfactory start-up can occur. This is based on the criterion that burn-out
does not occur, i.e. the saturation pressure in the heated zones should not
exceed the maximum capillary force. If burn-out is allowed to occur, drying
of the wick results, inhibiting the return flow of liquid.

A relationship which gives the maximum allowable heat input rate during the
start-up condition is:

$$Q_{max} = 0.4 \ \pi r_c^2 \times 0.73 \ L \ (P_E \ \rho_E)^{\frac{1}{2}} \qquad \ldots 3.10$$

where r_c is the vapour channel radius,

        L   is the latent heat of vaporisation

and P_E, ρ_E is the vapour pressure and density in the evaporator
             section.

It is important to meet start-up criteria when a heat pipe is used in an
application which may involve numerous starting and stopping actions, for
example in cooling a piece of electronic equipment or cooling brakes. One
way in which the problem can be overcome is to use an extra heat source
connected to a small branch heat pipe when the primary role of cooling is
required, thus reducing the number of start-up operations. The start-up time
of gas buffered heat pipes (see Chapter 6) is quicker.

Recently Busse (3.42) has made a significant contribution to the analysis of
the performance of heat pipes, showing that before sonic choking occurs, a
viscous limitation which can lie well below the sonic limit can be met. This
is described in detail in Chapter 2.

Sample Design Calculation A.

A heat pipe is required which will be capable of transferring a minimum of 15 W at vapour temperatures between 0 and 80°C over a distance of 1 m in zero gravity (a satellite application). Restraints on the design are such that the evaporator and condenser sections are each 8 cm long, located at each end of the heat pipe, and the maximum permissible temperature drop between the outside wall of the evaporator and the outside wall of the condenser is 6°C. Because of weight and volume limitations, the cross-sectional area of the vapour space should not exceed 0.197 cm². The heat pipe must also with-stand bonding temperatures.

Design a heat pipe to meet this specification.

1. Selection of Materials and Working Fluid

The selection of wick and wall materials is based on the criteria discussed in in this Chapter, with mass being an important parameter.

Aluminium alloy 6061 (HT30) is chosen for the wall, and stainless steel for the wick.

Working fluids compatible with these materials, based on available data, include:

> Freon 11 (UK equivalent - Arcton 11)
> Freon 113 (UK equivalent - Arcton 113)
> Acetone
> Ammonia

The limitations on heat transport must now be examined for each working fluid.

Sonic limit: The minimum axial heat flux due to the sonic limitation will occur at the minimum operating temperature, 0°C, and can be calculated from the equation:

$$q_s = \rho_v L \sqrt{\frac{\gamma \tilde{R} T_v}{2(\gamma + 1)m}}$$

All fluid properties are evaluated at $0^{\circ}C$, and the following values of ϕ_s are obtained:

Freon 11	0.69 kW/cm^2
Freon 113	3.1 kW/cm^2
Acetone	1.3 kW/cm^2
Ammonia	86 kW/cm^2

The limiting heat fluxes are obviously much greater than the requirement ($\frac{15}{0.197}$ W/cm^2, i.e. 76 W/cm^2) and therefore the sonic limit will not be encountered using any of the above fluids.

Entrainment limit: The maximum heat transport due to the entrainment limit may be determined using the equation:

$$Q_{ent} = \pi r_v^2 L \sqrt{\frac{2\pi \, \rho_v \, \sigma_\ell \, \cos\theta}{\lambda}}$$

λ is the characteristic dimension of the liquid/vapour interface, and for fine mesh may be taken as 0.036 mm.

Evaluating for acetone at $80^{\circ}C$

$$L = 495 \frac{kJ}{kg}$$

$$\sigma_\ell = 0.0162 \text{ N/m}$$

$$\rho_v = 4.05 \text{ kg/m}^3$$

$$r_v = 2.5 \text{ mm}$$

$$\therefore Q = \pi \times 0.25^2 \times 495 \sqrt{\frac{2\pi \times 4.05 \times 0.0162 \times 1 \times 10^2}{0.0036}}$$

Dimensions: $cm^2 \frac{kJ}{kg} \sqrt{\frac{kg}{m^3} \times \frac{N}{m} \times \frac{1}{m}}$

$$1N = \frac{1m \times kg}{sec^2}$$

∴ term in square root is: $\dfrac{kg}{m^3} \times \dfrac{m \times kg}{sec^2 \times m} \times \dfrac{1}{m}$

or: $\dfrac{kg^2}{m^4 \ sec^2}$

Evaluating the square root term

$$Q = \pi \times 0.25^2 \times 495 \times 1.07 \times 10^2$$

Dimensions: $cm^2 \times \dfrac{kJ}{kg} \times \dfrac{kg}{m^2 \ sec}$

$1 \ kW \ hr = 1 \ kJ \times 2.778 \times 10^{-4}$

$1 \ m^2 = 1 \ cm^2 \times 10^{-4}$

$1 \ kW \ sec = 1 \ kJ \times 2.778 \times 10^{-4} \times 3.6 \times 10^3$

∴ $Q = \pi \times (0.25^2 \times 10^{-4}) \times (495 \times 2.778 \times 10^{-4} \times 3.6 \times 10^3)$
$\times (1.07 \times 10^2)$

Dimensions: $(m^2) \ (\dfrac{kW \ sec}{kg}) \ (\dfrac{kg}{m^2 \ sec})$

This simplifies to give Q in kW.

Multiplying out:

$$Q = 0.104 \ kW$$

$$= \underline{104 \ W}$$

∴ with acetone the entrainment limit is well above the required heat transport capability. If the calculations are repeated for the other working fluids, similar conclusions can be drawn.

Wicking limit: At this stage the wick form has still to be specified, but a qualitative comparison of the potential of the four fluids can be obtained by evaluating the Merit number, $\rho_\ell \dfrac{\sigma_\ell L}{\mu_\ell}$ over the vapour temperature range. This is shown in Fig. 3.6, and reveals the advantage of ammonia, and to a lesser extent acetone, over the Freons.

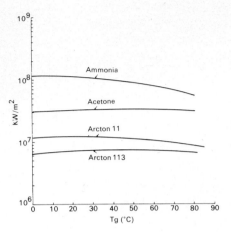

Fig. 3.6   Merit number for selected working fluids

Radial heat flux: Boiling in the wick may result in the vapour blocking the
supply of liquid to all parts of the evaporator. In arterial heat pipes
bubbles in the artery itself can create even more serious problems (see
Chapter 6). It is therefore desirable to have a working fluid with a high
superheat ΔT to reduce the chance of nucleation. The degree of superheat to
cause nucleation is given by:

$$\Delta T_S = \frac{3.06 \, T_v \, \sigma_\ell}{L \, \rho_v \, \delta}$$

where δ is the thermal layer thickness, and for this
                  application is taken as 0.15 mm (3.43). T_S is
                  evaluated at $80^\circ C$, as the lowest permissible
                  degree of superheat will occur at the maximum
                  operating temperature. These are (ΔT_S in $^\circ C$)

Freon 11	0.025
Freon 113	0.04
Acetone	0.58
Ammonia	0.002

These figures suggest that the Freons and ammonia require only very small
superheat temperatures at $80^\circ C$ to cause boiling. Acetone is the best fluid
from this point of view.

Priming of the wick: A further factor in fluid selection is priming ability (see Chapter 3). A comparison of the priming ability of fluids may be obtained from the ratio σ_ℓ/ρ_ℓ , and this is plotted against vapour temperature in Fig. 3.7 .

Acetone and ammonia are shown to be superior to the Freons over the whole operating temperature range.

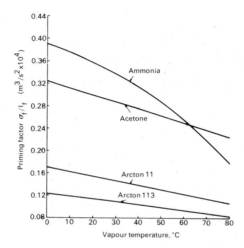

Fig. 3.7   Priming factor for selected fluids

Wall thickness: The requirement of this heat pipe necessitates the ability to be bonded to a radiator plate. Depending on the type of bonding used, the heat pipe may reach 170°C during bonding, and therefore vapour pressure is important in determining the wall thickness.

At this temperature the vapour pressures of ammonia and acetone are 113 and 17 bar respectively. Taking the 0.1% proof stress of HT30 aluminium as 46.3 MN/m^2 (allowing for some degradation of properties in weld regions), and using the thin cylinder formula, the minimum wall thickness for ammonia is 0.65 mm and 0.1 mm for acetone. There is therefore a mass penalty attached to the use of ammonia.

Conclusions on selection of working fluid: Acetone and ammonia both meet the heat transport requirements, ammonia being superior to acetone in this

respect. Nucleation occurs more readily in an ammonia heat pipe, and the
pipe may also be heavier. The handling of ammonia to obtain high purity
(see Chapter 4) is difficult. Acetone is therefore selected in spite of the
inferior performance.

2. Detail Design

Two types of wick structure are proposed for this heat pipe, homogeneous and
arterial types. A homogeneous wick may be a mesh, twill or felt, and arterial
types normally incorporate a mesh to distribute liquid circumferentially.
Homogeneous meshes are easy to form but have inferior properties to arterial
types. The first question is, therefore: Will a homogeneous wick transport
the required amount of fluid over 1 m to meet the heat transport specification?

To determine the minimum flow area to transport 15 W, one can equate the
maximum capillary pressure to the sum of the liquid and gravitational pressure
drops (neglecting vapour ΔP).

$$\Delta P_\ell + \Delta P_g = \Delta P_c$$

$$\text{where} \quad \Delta P_c \quad = \quad \frac{2\sigma_\ell \cos\theta}{r_c}$$

$$\Delta P_g \quad = \quad \rho_\ell \, g \, \ell \, \sin\phi$$

$$\Delta P_\ell \quad = \quad \frac{\mu_\ell}{\rho_\ell \, L} \times \frac{Q\ell_{eff}}{A_w \, K}$$

Taking properties at 80°C and assuming a mesh size of 400, and assuming

$$\ell \sin\phi = \quad 1 \text{ cm (end to end tilt plus tube diameter)}$$

$$\ell_{eff} \quad = \quad 100 \text{ cm}$$

$$\cos\theta \quad = \quad 1$$

and calculating K using the Blake-Koseny equation

$$K = \frac{d_w^2 (1 - \varepsilon)^3}{66.6\ \varepsilon^2}$$

where ε is the volume fraction of the solid phase (0.314)

and d_w is the wire diameter (0.025 mm)

We obtain $\Delta P_\ell = \dfrac{0.192 \times 15 \times 1}{719 \times 495 \times 0.3 \times 10^{-10} A_w}$

Dimensions: $\underline{(\text{Centipoise})(W)(m)}$

$(\dfrac{kg}{m^3})\ (\dfrac{kJ}{kg})\ (m^2)\ (m^2)$

Centipoise $\times$ 3.6 = kg/m hr

kJ $\times$ 2.778 $\times$ 10^{-4} = kW hr

$\therefore \Delta P_\ell = \dfrac{0.192 \times 3.6 \times 15 \times 10^{-3} \times 1}{719 \times 495 \times 2.778 \times 10^{-4} \times 0.3 \times 10^{-10} A_w}\quad \dfrac{kg}{m\ hr^2}$

$= \dfrac{0.192 \times 3.6 \times 15 \times 10^{-3}}{719 \times 495 \times 2.778 \times 10^{-4} \times 0.3 \times 10^{-10} \times 3600^2\ A_w}\quad \dfrac{N}{m^2}$

$= \dfrac{0.027}{A_w}\quad \dfrac{N}{m^2}$

$\Delta P_c = \dfrac{2 \times 0.0162 \times 1}{0.0029 \times 10^{-2}}$

Dimensions: $\dfrac{N}{m}\ /m$

$\therefore \Delta P_c = 1120\ \dfrac{N}{m^2}$

$\Delta P_g = 719 \times 9.81 \times 0.01$

Dimensions: $(\frac{kg}{m^3})$ $(\frac{m}{sec^2})$ (m)

$$\Delta P_g = 70 \frac{N}{m^2}$$

Equating the three ΔP terms, we obtain:

$$\frac{0.027}{A_w} + 70 = 1120$$

Rearranging:

$$A_w = \frac{0.027}{1120 - 70}$$

$$= 0.026 \times 10^{-3} m^2$$

$$= \underline{0.26 \ cm^2}$$

It can therefore be concluded that the homogeneous type of wick will not be acceptable, as the area required (0.26 cm^2) is greater than the total vapour space available (0.197 cm^2). An arterial wick must therefore be used.

Arterial diameter: Equation 3.9 in Chapter 3 describes the artery priming capability, setting a maximum value on the size of any arteries:

$$d_a = \frac{1}{2}\left[(\sqrt{h^2 + \frac{8\sigma_\ell \cos\theta}{(\rho_\ell - \rho_v)g}}) - h\right]$$

Using this equation and evaluating d_a at a vapour temperature of 30°C (for convenience priming ability may be demonstrated at room temperature), the maximum permitted value is 0.58 mm. To allow for uncertainties in fluid properties, wetting (θ assumed 0°C) and manufacturing tolerances, a practical limit is 0.5 mm. In the equation h is taken as 1 cm to cater for arteries near the top of the vapour space.

Circumferential liquid distribution; The circumferential wick thickness is limited by the fact that the temperature drop between the vapour space and the outside surface of the heat pipe and vice versa should be 3°C maximum. Assuming that the temperature drop through the aluminium wall is negligible,

the thermal conductivity of the wick may be determined and used in the
steady state conduction equation.

$$k_{wick} = (\frac{\beta - \epsilon}{\beta + \epsilon}) k_\ell$$

where $\beta = (1 + \frac{k_s}{k_\ell})/(1 - \frac{k_s}{k_\ell})$

$k_s = 16$ W/m $^\circ$C (steel)

$k_\ell = 0.165$ W/m $^\circ$C (acetone)

$$\therefore \beta = \frac{1 + 97}{1 - 97} = -1.02$$

The volume fraction ϵ of the solid phase is approximately 0.3

$$\therefore k_{wick} = (\frac{-1.02 - 0.3}{-1.02 + 0.3}) \times 0.165$$

$$= 0.3 \text{ W/m } ^\circ C$$

Using the basic conduction equation:

$$Q = kA \frac{dT}{dx}$$

$$dx = k \, dT \frac{A}{Q}$$

$$= 0.3 \times 3 \times \frac{10^{-4}}{1.2}$$

Dimensions: $\frac{W}{m \, ^\circ C} \times ^\circ C \times \frac{m^2}{W}$

$$= 0.075 \times 10^{-3} \text{ m}$$

$$= \underline{0.075 \text{ mm}}$$

Thus the circumferential wick must be 400 mesh, which has a thickness of
0.05 mm. Coarser meshes are too thick.

Arterial wick: Returning to the artery, the penultimate section revealed
that the maximum artery depth permissible was 0.5 mm. In order to prevent
nucleation in the arteries, they should be kept away from the heat pipe wall,
and formed of low conductivity material. It is also necessary to cover the
arteries with a fine pore structure, and 400 mesh stainless steel is selected.

It is desirable to have several arteries to give a degree of redundancy, and
two proposed configurations are considered, one having six arteries as shown
in Fig. 3.3, and the other four arteries. In the former case each groove is
nominally 1.0 mm wide, in the second case 1.5 mm.

It is now possible to predict the overall capability of the heat pipe, to
check that it meets the specification.

We have already shown that entrainment and sonic limitations will not be met,
and that the radial heat flux is acceptable. The heat pipe should also meet
the overall temperature drop requirement, and the arteries are sufficiently
small to allow repriming at 30°C. The wall thickness requirement for
structural integrity (0.1 mm minimum) can easily be satisfied. The wicking
limitation will therefore determine the maximum performance.

Final analysis: The wicking limitation is reached when the liquid, vapour
and gravitational pressure losses are equal to the maximum capillary pressure.

$$\text{i.e.} \quad \Delta P_{\ell a} + \Delta P_{\ell m} + \Delta P_g + \Delta P_v = \Delta P_c$$

where $\Delta P_{\ell a}$ is the pressure drop in the artery

$\Delta P_{\ell m}$ is the loss in the circumferential wick

The axial flow in the mesh will have little effect and can be neglected.

McAdams (3.44) presents an equation for the pressure loss, assuming laminar
flow, in a rectangular duct, and shows that the equation is in good agreement
with experiment for streamline flow in rectangular ducts having depth/width
ratios ($\frac{a_a}{b_a}$) of 0.05 - 1.0.

This equation may be written:

$$\Delta P_{\ell a} = \frac{4 K_\ell \times \ell_{eff} \quad Q}{a_a^2 \, b_a^2 \quad \phi_c \quad N}$$

where   N   is the number of channels

ϕ_c is a function of channel aspect ratio and is given
in Fig. 3.8

and $K_\ell \quad = \quad \dfrac{\mu_\ell}{\rho_\ell L}$

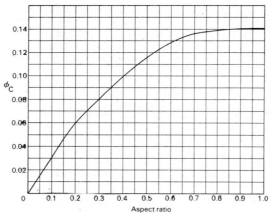

Fig. 3.8   Channel aspect ratio factor

The summed pressure loss in the condenser and evaporator is given by:

$$\Delta P_{\ell m} = \frac{K_\ell \times \ell_{effc} \quad Q}{2 \, K \, A_c}$$

where ℓ_{effc} is the effective circumferential flow length
(approximately $\frac{\pi r_W}{4}$) carrying $\frac{\dot{m}}{4}$

where $\dot{m}$ is the liquid mass flow.

A_c is the circumferential flow area (mesh thickness
× cond. or evap. length)

K is the permeability of 400 mesh

The circumferential flow area for 400 mesh with 2 layers is:

$$A_c \; = 8 \times 10^{-2} \times 0.1 \times 10^{-3} \; m^2$$

$$= 8 \times 10^{-6} \; m^2$$

A resistance occurs in both the evaporator and condenser, therefore, substituting in the above equation

$$\Delta P_{\ell m} = \frac{\pi \times 2.5 \times 10^{-3}}{2 \times 4} \quad \frac{1}{8 \times 10^{-6}} \quad \frac{K_\ell \, Q}{0.314 \times 10^{-10}}$$

$$\approx \underline{4.00 \times 10^{12} \; K_\ell \, Q} \qquad \text{for each section}$$

$$\Delta P_{\ell a} = \frac{4 \times 0.92 \times K_\ell \times Q \times 10^{12}}{(0.5)^2 (1)^2 \times 0.115 \times 6}$$

$$= \underline{21.3 \times 10^{12} \; K_\ell \, Q} \qquad \text{for 6 channels}$$

$$= \frac{4 \times 0.92 \times K_\ell \, Q \times 10^{12}}{(0.5)^2 (1.5)^2 \times 0.088 \times 4}$$

$$= \underline{18.59 \times 10^{12} \; K_\ell \, Q} \qquad \text{for 4 channels}$$

The vapour pressure loss, which occurs in two near-semicircular ducts, can be obtained using the Hagen-Poiseuille equation if the hydraulic radius is used:

$$\Delta P_v \; = \frac{1}{2} \left\{ \frac{8 K_v \, \ell_{eff} \, Q}{\pi r_H^4} \right\}$$

where $K_v = \dfrac{\mu_v}{\rho_v L}$

Now the axial Reynolds number Re_z is given by:

$$Re_z = \frac{Q}{\pi r_H \mu_\ell L}$$

The transitional heat load can be calculated assuming that transition from laminar to turbulent flow occurs at $Re_z = 1000$, based on hydraulic radius. Restricting the width of the stainless steel former in Fig. 3.3 to 1.5 mm, $r_H = 1.07$ mm.

Q may be evaluated at the transition point using values of μ_v and L at several temperatures between 0 and 80°C, giving:

Vapour Temp. ($^\circ$C)	Transitional Load (W)
0	31.1
20	31.2
40	30.6
60	30.2
80	30.0

The transitional load is always greater than the design load of 15 W, but as the heat pipe may be capable of operating in excess of the design load, it is necessary to investigate the turbulent regime.

For $Re_z > 1000$ and for 2 ducts:

$$\Delta P_v = \frac{0.00896 \; \mu_v^{0.25} \; Q^{1.75} \; \ell_{eff}}{2 \rho_v \; r_H^{4.75} \; L^{1.75}}$$

This is the empirical Blasius equation.

We can now evaluate ΔP_v for laminar and turbulent-flow:

$$\Delta P_v \text{ (laminar)} = \frac{1}{2} \left\{ \frac{8 \times 0.92 \ K_v \times Q \times 10^{12}}{\pi \times 1.07^4} \right\}$$

$$\approx \underline{0.9 \times 10^{12} \ K_v \times Q}$$

$$\Delta P_v \text{ (turb)} = \frac{0.00896 \times 0.92}{2 \times (1.07 \times 10^{-3})^{4.75}} \ Q^{1.75} \left(\frac{\mu_v}{L^{1.75} \rho_v}\right)^{0.25}$$

$$= \underline{0.53 \times 10^{12} \ Q^{1.75} \ \left(\frac{\mu_v}{L^{1.75} \rho_v}\right)^{0.25}}$$

The gravitational pressure drop is:

$$\Delta P_g = \rho_\ell \ g \ \ell \ \sin \phi$$

$$= \underline{0.0981 \ \rho_\ell ,}$$

taking $\ell \sin \phi = 1 \text{ cm}$

The capillary pressure generated by the arteries is given by:

$$\Delta P_c = \frac{2\sigma_\ell \cos \theta}{r_c}$$

$$= \frac{2}{0.003 \times 10^{-2}} \ \sigma_\ell \cos \theta$$

$$= \underline{0.667 \times 10^5 \ \sigma_\ell}$$

Summarising:

$$\Delta P_c = 0.667 \times 10^5 \, \sigma_\ell$$

$$\Delta P_g = 0.0981 \, \rho_\ell$$

$$\Delta P v_\ell = 0.9 \times 10^{12} \, K_v Q$$

$$\Delta P_{v_t} = 0.53 \times 10^{12} \, Q^{1.75} \left(\frac{\mu_v^{0.25}}{L^{1.75} \, \rho_v}\right)$$

$$\Delta P_{\ell_m} = 4 \times 10^{12} \, K_\ell Q$$

$$\Delta P_{\ell_a} = 21.3 \times 10^{12} \, K_\ell Q \qquad \text{(6 channels)}$$

$$= 18.59 \quad 10^{12} \, K_\ell Q \qquad \text{(4 channels)}$$

These equations involve Q and the properties of the working fluid. Using properties at each temperature (in 20°C increments) over the operating range, the total capability can be determined:

$$\Delta P_c = \Delta P_{\ell_m} + \Delta P_{\ell_a} \left\{ \begin{array}{l} \text{6 channels} \\ \text{4 channels} \end{array} \right\} + \Delta P_g + \Delta P_v \left\{ \begin{array}{l} \text{laminar} \\ \text{turbulent} \end{array} \right\}$$

This yields the following results:

Vapour Temp ($^\circ$C)	Q(W)			
	Laminar 4 channels	Laminar 6 channels	Turbulent 4 channels	Turbulent 6 channels
0	21.6	20.9	–	–
20	34.0	32.5	22.6	22.0
40	42.6	40.2	27.9	27.0
60	49.1	45.8	33.0	32.0
80	51.4	47.6	36.4	35.0

This heat pipe was constructed with 6 grooves in the artery structure, and met the specification.

The stages in the design may now be listed:

(i)   Select wick and wall materials

(ii)   Select working fluid(s):
            Criteria - limitations
                          pressure
                          priming
                          handling
                          purity etc.

(iii)   Examine wick types:
            Homogeneous rejected
            Arterial selected

(iv)   Determine artery sizes

(v)   Examine radial resistance to heat flow

(vi)   Examine overall pressure balance of proposed design

(vii)   Select final configuration on basis of (vi) and such
           features as manufacturing difficulties etc.

Sample Design Calculation B

Problem: Obtain an approximation to the liquid flow rate and heat transport
capability of a simple water heat pipe operating at $100^{\circ}C$ having a wick of
2 layers of 250 mesh against the inside wall. The heat pipe is 30 cm long and
has a bore of 1 cm diameter. It is operating at an inclination to the
horizontal of 30° , with evaporator above the condenser.

The maximum heat transport in a heat pipe at a given vapour temperature may
be obtained from the equation:

$$Q_{max} = \dot{m}_{max} \, L$$

where $\dot{m}_{max}$ is the maximum liquid flow rate in the wick

Using the standard pressure balance equation:

$$\Delta P_c \;=\; \Delta P_v + \Delta P_\ell + \Delta P_g$$

and neglecting, for the purposes of a first approximation, the vapour pressure drop ΔP_v, we can substitute for the pressure terms and obtain:

$$\frac{2\sigma_\ell \cos\theta}{r_c} \;=\; \frac{\mu_\ell}{\rho_\ell \, L} \times \frac{Q \, \ell_{eff}}{A_w \, K} + \rho_\ell \, g \, \ell \, \sin\phi$$

Rearranging and substituting for $\dot{m}$, we obtain:

$$\dot{m} \;=\; \frac{\rho_\ell \, K \, A_w}{\mu_\ell \, \ell_{eff}} \left\{ \frac{2\,\sigma_\ell}{r_c} \cos\theta - \rho_\ell \, g \, \ell_{eff} \sin\phi \right\}$$

The wire diameter of 250 mesh is typically 0.0045 cm, and therefore the thickness of 2 layers of 250 mesh is 4×0.0045 cm or 0.0180 cm.

The bore of the heat pipe is 1 cm,

$$\therefore A_{wick} \simeq 0.018 \times \pi \times 1$$

$$\simeq \underline{0.057 \text{ cm}^2}$$

From Table 3.3 the pore radius r_c and permeability K of 250 mesh are 0.002 cm and 0.302×10^{-10} m^2 respectively.

Assuming perfect wetting ($\theta = 0^\circ$), the mass flow $\dot{m}$ may be calculated using the properties of water at 100°C.

$$L \;=\; 2.258 \times 10^6 \text{ J/kg}$$

$$\rho_\ell \;=\; 958 \text{ kg/m}^3$$

$$\mu_\ell \;=\; 0.283 \text{ mNs /m}^2$$

$$\sigma_\ell \;=\; 58.85 \text{ mN/m}$$

$$\therefore \dot{m}_{max} = \frac{958 \times 0.302 \times 10^{-10} \times 0.057 \times 10^{-4}}{0.283 \times 10^{-3} \times 0.3} \left\{ \frac{2 \times 58.85 \times 10^{-3}}{0.2 \times 10^{-4}} \right.$$

$$\left. -958 \times 9.810 \times 0.3 \times 0.5 \right\}$$

Dimensions:
$$\frac{kg}{m^3} \times m^2 \times m^2 \times \frac{m^2}{Ns} \frac{1}{m} \left\{ \frac{N}{m} \frac{1}{m} - \frac{kg}{m^3} \times \frac{m}{s^2} \times m \right\}$$

[Note: $1N = \dfrac{1 \ kg \times m}{s^2}$]

Dimensions: $\dfrac{kg \times m^2}{Ns} \left\{ \dfrac{N}{m^2} - \dfrac{N}{m^2} \right\} = \dfrac{kg}{s}$ - correct

Multiplying out $\dot{m}_{max} = 1.95 \times 10^{-9} (5885 - 1410)$

$$\therefore \underline{\dot{m}_{max} = 8.636 \times 10^{-6} \ kg/s}$$

Now $Q_{max} = \dot{m}_{max} \times L$

$$\therefore \ Q_{max} = 8.636 \times 10^{-6} \times 2.258 \times 10^{6}$$

Dimensions: $\left(\dfrac{kg}{s} \times \dfrac{J}{kg} \right) = \dfrac{J}{s} = W$

$$\therefore \quad Q_{max} = \underline{19.5 \ W}$$

Sample Design Calculation C

Problem: The capability of the above heat pipe is low. What improvement will be made if 2 layers of 100 mesh are added to the 250 mesh wick, to increase liquid flow capability.

The wire diameter of 100 mesh is 0.010 cm

$\therefore$ thickness of 2 layers is 0.040 cm

Total wick thickness = 0.040 + 0.018 cm

$$= \underline{0.058 \text{ cm}}$$

$\therefore A_{wick}$ $\approx$ $0.058 \times \pi \times 1$

$\approx$ $\underline{0.182 \text{ cm}^2}$

The capillary pressure is still governed by the 250 mesh, and r_c = 0.002 cm.

The permeability of 100 mesh is used, Langston and Kunz giving a value of $1.52 \times 10^{-10} \text{m}^2$.

The mass flow may now be calculated:

$$\dot{m}_{max} = \frac{958 \times 1.52 \times 10^{-10} \times 0.182 \times 10^{-4}}{0.283 \times 10^{-3} \times 0.3} \{5885 - 1410\}$$

$$= 3.1 \times 10^{-8} \, (4.475 \times 10^3)$$

$$\dot{m}_{max} = 1.39 \times 10^4 \text{ kg/sec}$$

This is over one order of magnitude greater than that given by the 250 mesh alone.

$$Q_{max} = 1.39 \times 10^{-4} \times 2.258 \times 10^6$$

$$= \underline{314 \text{ W}}$$

HP—F

152   P.D. DUNN   D.A. REAY

REFERENCES

3.1   Asselman, G.A.A. and Green, D.B. The heat pipe - an applied technology. Philips Technical Review, 1972 pp 32 - 34.

3.2   Deverall, J.E. Mercury as a heat pipe fluid. ASME Paper 70-HT/SpT-8, American Society of Mechanical Engineers, 1970.

3.3   Reay, D.A. Mercury wetting of wicks. Proc. 4th C.H.I.S.A. Conf., Prague Sept. 1972.

3.4   Bienert, W. Heat pipes for solar energy collectors. 1st International Heat Pipe Conf. Stuttgart, Paper 12-1, Oct. 1973.

3.5   Quataert, D., Busse, C.A. and Geiger, F. Long time behaviour of high temperature tungsten-rhenium heat pipes with lithium or silver as the working fluid. Paper 4-4. ibid.

3.6   Lidbury, J.A. A helium heat pipe. Nimrod Design Group Report NDG-72-11, Rutherford Laboratory, England, 1972.

3.7   Groll, M. Wärmerohre als Baudemente in der Wärme-und Kältetechnik. Brennst-Waermekraft. Vol. 25, No. 1, Jan. 1973 (German).

3.8   Marto, P.J. and Mosteller, W.L. Effect of nucleate boiling on the operation of low temperature heat pipes. ASME Paper 69-HT-24. New York, 1969.

3.9   Phillips, E.C. Low Temperature heat pipe research program. NASA CR-66792, 1970.

3.10 Keser, D. Experimental determination of properties of saturated sintered wicks. 1st International Heat Pipe Conf. Stuttgart, 1973.

3.11 Moritz, K. and Pruschek, R. Limits of energy transport in heat pipes. Chemie Ing. Technik. Vol. 41, No. 1, 2, 1969 (German).

3.12 Vinz, P. and Busse, C.A. Axial heat transfer limits of cylindrical sodium heat pipes between 25 W/cm^2 and 15.5 kW/cm^2. 1st International Heat Pipe Conf. Paper 2-1, Stuttgart, 1973.

3.13 Busse, C.A. Heat pipe research in Europe. Euratom Report. EUR 4210 f, 1969.

3.14 Quataert, D, Busse, C.A. and Geiger, F. Long term behaviour of high temperature tungsten-rhenium heat pipes with lithium or silver as working fluid. 1st International Heat Pipe Conf., Paper 4-4, Stuttgart, 1973.

3.15 Schroff, A.M. and Armand, M. Le Caloduc. Revue Technique Thompson -CSF, Vol. 1, No. 4, Dec. 1969. (French).

3.16 Farran, R.A. and Starner, K.E. Determining wicking properties of compressible materials for heat pipe applications. Proc. Avn. & Space Conf., Beverly Hills, California. June 1968. ASME pp 659 - 70.

3.17 Ferrell, J.K. and Alleavitch, J. Vaporisation heat transfer in capillary wick structures. Dept. Chem. Engng. Report, North Carolina University, Raleigh, USA, 1969.

3.18   Freggens, R.A. Experimental determination of wick properties for
        heat pipe applications. 4th Intersoc. Energy Conversion Engng.
        Conf. Washington D.C., 22 - 26 Sept. 1969, pp 888 - 897.

3.19   Phillips, E.C. and Hinderman, J.D. Determination of properties of
        capillary media useful in heat pipe design. ASME Paper 69-HT-18,
        1969.

3.20   Birnbreier, H. and Gammel, G. Measurement of the effective
        capillary radius and the permeability of different capillary
        structures. 1st International Heat Pipe Conf., Paper 5-4,
        Stuttgart, Oct. 1973.

3.21   Langston, L.S. and Kunz, H.R. Liquid transport properties of some
        heat pipe wicking materials. ASME Paper 69-HT-17, 1969.

3.22   McKinney, B.G. An experimental and analytical study of water heat
        pipes for moderate temperature ranges. NASA-TM-X-53849. Alabama:
        Marshall Space Flight Center, June, 1969.

3.23   Calimbas, A.T. and Hulett, R.H. An avionic heat pipe. ASME Paper
        69-HT-16. New York, 1969.

3.24   Katzoff, S. Heat pipes and vapour chambers for thermal control of
        spacecraft. AIAA Paper 67-310. 1967.

3.25   Hoogendoorn, C.J. and Nio, S.G. Permeability studies on wire
        screens and grooves. 1st International Heat Pipe Conf., Paper 5-3,
        Stuttgart, Oct. 1973.

3.26   Chun, K.R. Some experiments on screen wick dry-out limits. ASME
        Paper 71-WA/HT-6, 1971.

3.27   Ivanovskii, M.N. et al. Investigation of heat and mass transfer in
        a heat pipe with a sodium coolant. High Temp. Vol. 8, No. 2,
        pp 299 - 304, 1970.

3.28   Gorring, R.L. and Churchill, S.W. Thermal conductivity of hetero-
        geneous materials. Chem. Engng. Progress. Vol. 57, No. 7, July
        1961.

3.29   Maxwell, J.C. A treatise on electricity and magnetism. Vol. 1,
        3rd Edn. OUP (1891), reprinted by Dover, New York, 1954.

3.30   Joy, P. Optimum cryogenic heat pipe design. ASME Paper 70-HT/SpT
        -7, 1970.

3.31   Eggers, P.E. and Serkiz, A.W. Development of cryogenic heat pipes.
        ASME Paper 70-WA/Ener-1, New York, 1970.

3.32   Kosowski, N. and Kosson, R. Experimental performance of grooved
        heat pipes at moderate temperatures. AIAA Paper 71-409, 1971.

3.33   Basiulis, A. and Filler, M. Operating characteristics and long
        life capabilities of organic fluid heat pipes. AIAA Paper 71-408.
        6th AIAA Thermophys. Conf., Tullahoma, Tennessee, April, 1971.

3.34   Kreeb, H., Groll, M. and Zimmermann, P. Life test investigations
        with low temperature heat pipes. 1st International Heat Pipe Conf.
        Stuttgart, Paper 4-1. Oct. 1973.

3.35   Busse, C.A., Campanile, A. and Loens, J. Hydrogen generation in
        water heat pipes at 250°C. 1st International Heat Pipe Conf.,
        Stuttgart, Paper 4-2, Oct. 1973.

3.36   Gerrels, E.E. and Larson, J.W. Brayton cycle vapour chamber (heat
        pipe) radiator study.. NASA CR-1677, General Electric Company,
        Philadelphia, NASA. Feb. 1971.

3.37   Dix, E.H. et al. The resistance of aluminium alloys to corrosion.
        Metals Handbook, Vol. 50, American Society for Metals, Ohio, 1961.

3.38   Freon T.F. Solvent, E.I. Dupont de Nemours & Co. Inc. Tech.
        Bulletin FSR-1, 1965.

3.39   Marcus, B.D. Theory and design of variable conductance heat pipes.
        TRW Systems Group, NASA CR-2018, April, 1972.

3.40   Kemme, J.E. Heat pipe capability experiments. Los Alamos Scien-
        tific Laboratory. Report LA-3585. Aug. 1966.

3.41   Van Andel, E. Heat pipe design theory. Euratom Center for
        Information and Documentation. Report EUR No. 4210 e,f, 1969.

3.42   Busse, C.A. Theory of the ultimate heat transfer limit of
        cylindrical heat pipes. Int. Journal of Heat and Mass Transfer,
        Vol. 16, pp 169 - 186, 1973.

3.43   Weibe, J.R. and Judd, R.L. Superheat layer thickness measurements
        in saturated and sub-cooled nucleate boiling. Trans. ASME, Journ.
        Heat Transfer, Nov. 1971.

3.44   McAdams, W.H. Heat transmission. 3rd Edn. McGraw-Hill Book Co.
        Inc. 1954.

3.45   Basiulis, A., Prager, R.C. and Lamp, T.R. Compatibility and reli-
        ability of heat pipe materials. Proc. 2nd Int. Heat Pipe Confer-
        ence, Bologna: ESA Report SP 112, 1976.

3.46   Anderson, W.T. Hydrogen evolution in nickel-water heat pipes.
        AIAA Paper 73-726, AIAA Conf., Palm Springs, California, 1973.

3.47   Geiger, F. and Quataert, D. Corrosion studies of tungsten heat
        pipes at temperatures up to 2650°C. Proc. 2nd Int. Heat Pipe Con-
        ference, Bologna; ESA Report SP 112, 1976.

3.48   Polasek, F. and Stulc, P. Heat pipes for the temperature range
        200°C to 600°C. Ibid.

Heat Pipe Manufacture and Testing

The manufacture of heat pipes involves a number of comparatively simple
operations, particularly when the unit is designed for operation at temperatures of the order of, say 50 - 200°C. It embraces skills such as welding,
machining, chemical cleaning and non-destructive testing, and can be carried
out following a relatively small outlay on capital equipment. The most
expensive item is likely to be leak detection equipment.

With all heat pipes, however, cleanliness is of prime importance to ensure
that no incompatibilities exist, (assuming that the materials selected for
the wick, wall and working fluid are themselves compatible), and to make
certain that the wick and wall will be wetted by the working fluid. As well
as affecting the life of the heat pipe, negligence in assembly procedures can
lead to inferior performance, due, for example, to poor wetting. Atmospheric
contaminants, in addition to those likely to be present in the raw working
fluid, must be avoided. Above all the heat pipe must be leak-tight to a very
high degree. This can involve out-gassing of the metal used for the heat
pipe wall, end caps etc., although this is not essential for simple low temperature operations.

Quality control cannot be over-emphasised in heat pipe manufacture, and in
the following discussion of assembly methods, this will be frequently
stressed.

A substantial part of this Chapter is allocated to a review of life test
procedures for heat pipes. The life of a heat pipe often requires careful
assessment in view of the many factors which can affect long term performance,
and most establishments seriously involved in heat pipe design and
manufacture have extensive life test programmes in progress. As discussed
later, data available from the literature can indicate satisfactory wall/wick
/working fluid combinations, but the assembly procedures used differ from one
manufacturer to another, and this may introduce an unknown factor which will
necessitate investigation.

Measuring the performance of heat pipes is also a necessary part of the work
leading to an acceptable product, and the interpretation of the results may
prove difficult. Test procedures for heat pipes destined for use in orbiting
satellites have their own special requirements brought about by the need to
predict performance in zero gravity by testing in earth gravity.

4.1 Manufacture and Assembly

4.1.1 Container materials.

The heat pipe container, including the end caps
and filling tube, is selected on the basis of several properties of the
material used, and these are listed in Chapter 3. (Unless stated otherwise,
the discussion in this Chapter assumes that the heat pipes are tubular in
geometry). However, the practical implications of the selection are numerous.

Of the many materials available for the container, three are by far the most
common in use, namely copper, aluminium and stainless steel. Copper is
eminently satisfactory for heat pipes operating between 0 and 200°C in
applications such as electronics cooling. While commercially pure copper
tube is suitable, the oxygen-free high conductivity type (OFHC) is preferable.
Like aluminium and stainless steel, the material is readily available and can
be obtained in a wide variety of diameters and wall thicknesses in its
tubular form.

Aluminium is less common as a material in commercially available heat pipes,
but has received a great deal of attention in aerospace applications, because
of its obvious weight advantages. It is generally used in alloy form,
typically 6061-T6, the nearest British equivalent being aluminium alloy HT30.
Again this is readily available and can be drawn to suit by the heat pipe
manufacturer, or extruded to incorporate, for example, a grooved wick.

Stainless steel unfortunately cannot generally be used as a container
material with water where a long life is required, owing to gas generation
problems, but it is perfectly acceptable with many other working fluids, and
is in many cases the only suitable container, as for example, with liquid
metals such as mercury, sodium and potassium. Types of stainless steel
regularly used for heat pipes include 302, 316 and 321.

In the assembly of heat pipes provision must be made for filling, and the
most common procedure involves the use of an end cap with a small diameter
tube attached to it, as shown in Fig. 4.1. The other end of the heat pipe
contains a blank end cap. End cap and filling tube materials are generally
identical to those of the heat pipe case, although for convenience a copper
extension may be added to a stainless steel filling tube for cold welding
(see Section 4.1.9). It may be desirable to add a valve to the filling tube
where, for example, gas analysis may be carried out following life tests
(see Section 4.2). The valve material must, of course, be compatible with
the working fluid.

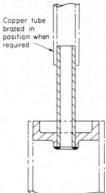

Copper tube
brazed in
position when
required

Fig. 4.1   End cap and filling tube

If the heat pipe is to operate at high vapour pressures, a pressure test
should be carried out to check the integrity of the vessel.

4.1.2 Wick materials and form.   The number and form of materials which
have been tested as wicks in heat pipes is very large. Reference has
already been made to some of these in analysis of the liquid pressure drop,
presented in Chapter 2, and in the discussion on selection criteria in
Chapter 3.

Wire mesh: The most common form of wick is a woven wire mesh or twill which
can be made in many metals. Stainless steel, monel and copper are woven to
produce meshes having very small pore sizes (see Table 3.3) and 400 mesh
stainless steel is available 'off the shelf' from several manufacturers.
(Appendix 4 lists manufacturers and stockists of many items of use to those

contemplating making heat pipes). Aluminium mesh is available, but because
of difficulties in producing and weaving fine aluminium wires, the requirements of small pore wicks cannot be met.

Stainless steel is the easiest material to handle in mesh form. It can be
rolled and retains its shape well. The inherent springiness in the coarse
meshes assist in retaining the wick against the heat pipe wall, in some cases
obviating the need for any other form of wick location. In heat pipes where
a 400 mesh is used, a coarse 100 mesh layer located at the inner radius can
hold the finer mesh in shape. Stainless steel can also be diffusion bonded
giving strong permanent wick structures attached to the heat pipe wall. The
diffusion bonding of stainless steel is best carried out in a vacuum furnace
at a temperature of 1150 - 1200°C.

The spot-welding of wicks is a convenient technique for preserving shape or
for attaching the wick to the wall in cases where the heat pipe diameter is
sufficiently large to permit insertion of an electrode. Failing this a coil
spring can be used.

It is important to ensure that whatever the wick form, it is in very close
contact with the heat pipe wall, particularly at the evaporator section,
otherwise local hot spots will occur. With mesh the best way of making
certain that this is the case is to diffusion bond the assembly.

Sintering: A similar structure having intimate contact with the heat pipe
wall is a sintered wick. Sintering is often used to produce metallic filters,
and many components of machines are now produced by this process as opposed
to die casting or moulding.

The process involves bonding together a large number of particles in the form
of a packed metal powder. The pore size of the wick thus formed can be
arranged to suit by selecting powders having a particular size. The powder,
which is normally spherical, is placed in containers giving the shape
required and then either sintered without being further compacted or, if a
temporary binder is used, a small amount of pressure may be applied. Sintering is normally carried out at a temperature 100 - 200°C below the melting
point of the sintering material.

The simplest way of making wicks by this method is to sinter the powder in the tube that will form the final heat pipe. This has the advantage that the wick is also sintered to the tube wall and thus makes a stronger structure. In order to leave the central vapour channel open a temporary mandrel has to be inserted in the tube. The powder is then placed in the annulus between mandrel and tube. In the case of copper powder a stainless steel mandrel is satisfactory as the copper will not bond to stainless steel and thus the bar can easily be removed after sintering. The bar is held in a central position at each end of the tube by a stainless steel collar.

A typical sintering process is described below. Copper was selected as the powder material and also as the heat pipe wall. The particle size chosen was -150 +300 grade, giving particles of 0.05 - 0.10 mm diameter. The tube was fitted with the mandrel and a collar at one end. The powder was then poured in from the other end. No attempt was made to compact the powder apart from tapping the tube to make sure there were no gross cavities left. When the tube was full the other collar was put in place and pushed up against the powder. The complete assembly was then sintered by heating in hydrogen at $850^{\circ}C$ for $\frac{1}{2}$ hour. After the tube was cool and removed from the furnace the mandrel was removed and the tube, without the mandrel, was then resintered. (The reason for this was that when the mandrel was in place the hydrogen could not flow easily through the powder and as a result sintering may not have been completely successful since the hydrogen is necessary to reduce the oxide film which hinders the process). After this operation the tube was ready for use. Fig. 4.2 shows a cross section of a completed tube and Fig. 4.3 shows a magnified view of the structure of the copper wick. The porosity of the finished wick is of the order of 40 - 50%.

A second type of sintering may be carried out to increase the porosity. This necessitates the incorporation of inert filler material to act as pore formers. This is subsequently removed during the sintering process, thus leaving a very porous structure. The filler used was a perspex powder which is available as small spheres. This powder was sieved to remove the -150 +300 (0.050 - 0.100 mm) fraction. This was mixed with an equal volume of very fine copper powder (-20 μm). On mixing, the copper uniformly coats the plastic spheres. This composite powder then shows no tendency to separate into its components.

Fig. 4.2 Sintered wick cross-section (Copper)
(Courtesy IRD)

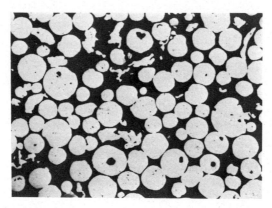

Fig. 4.3   Magnified view of sinter structure
(Courtesy IRD)

The wick is now made up exactly as for the previous tube with the exception
that more compaction is required in order to combat the very high shrinkage
that takes place during sintering. During the initial stages of the
sintering the plastic is vapourised and diffused out of the copper compact,
thus leaving a skeletal structure of fine copper powder with large interconnected pores. The final porosity is probably of the order of 75 - 85%.

It is obvious that there are many possible variations of the wicks made by
sintering methods. Porosity, capillary rise and volume flow can all be
optimised by the correct choice of metal powder size, filler size, filler

proportion and by incorporation of channel forming fillers.

Vapour deposition: Sintering is not the only technique whereby a porous
layer can be formed which is in intimate contact with the inner wall of the
heat pipe. Other processes include vapour-coating, cathode sputtering and
flame spraying. Brown Boveri, in UK Patent 1313525, describe a process known
as 'vapour-plating' which has been successfully used in heat pipe wick
construction. This involves plating the internal surface of the heat pipe
structure with a tungsten layer by reacting tungsten hexafluoride vapour with
hydrogen, the porosity of the layer being governed by the surface temperature,
nozzle movement, and distance of the nozzle from the surface to be coated.

Grooves: A type of wick which is widely used in spacecraft applications but
which is unable to support significant capillary heads in earth gravity, is a
grooved system. The simplest way of producing longitudinal grooves in the
wall of a heat pipe is by extrusion or by broaching. Aluminium is the most
satisfactory material for extruding, where grooves may be comparatively
narrow in width, but possess a greater depth. An example of a grooved heat
pipe is shown in Fig. 4.4. The external cross-section of the heat pipe can
also be adapted for a particular application. If the heat pipe is to be
mounted on a plate, a flat surface may be incorporated on the wall of the
heat pipe to give better thermal contact with the plate.

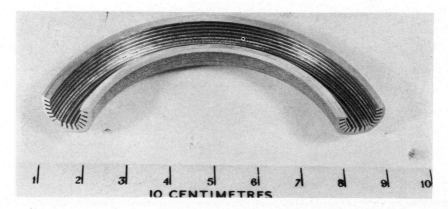

Fig. 4.4   Grooved wick (aluminium alloy) (Courtesy IRD)

An alternative groove arrangement involves 'threading' the inside wall of the
heat pipe using taps or a single point cutting tool to give a thread pitch of

up to 40 threads per cm. An example of this is shown in Fig. 4.5, in which
15 threads per cm. were formed on the inside of a 6 mm bore aluminium tube of
1 m length. These threaded arteries are attractive for circumferential
liquid distribution, and may be used in conjunction with a different artery
system for axial liquid transport.

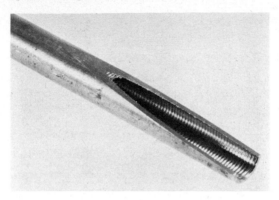

Fig. 4.5 Threaded artery in aluminium wall (Courtesy IRD)

Felts and foams: Several companies are now producing metal and ceramic felts
and metal foam which can be effectively used as heat pipe wicks, particularly
where units of non-circular cross section are required. The properties of
some of these materials are given in Table 3.3. Foams are available in
nickel and copper, and felt materials include stainless steel and woven
ceramic fibres (Refrasil). The names and addresses of manufacturers are given
in Appendix 4. Foams are available in sheet and rod form, and can be
supplied in a variety of pore sizes. Metallic felts are normally produced in
sheets, and are much more pliable than foams. An advantage of the felt is
that by using mandrels and applying a sintering process, longitudinal
arteries could be incorporated inside the structure, providing low resistance
flow paths.

Knitted ceramic fibres are available with very small pore sizes, and are
inert to most common working fluids. Because of their lack of rigidity,
particularly when saturated with a liquid, it is advisable to use them in
conjunction with a wire mesh wick to retain their shape and desired location.
The ceramic structure can be obtained in the form of multi-layer sleeves,
ideal for immediate use as a wick, and a range of diameters of sleeve is
available. Some stretching of the sleeve can be applied to reduce the

diameter should the exact size not be available.

4.1.3 Cleaning of container and wick. All of the materials used in a heat
pipe must be clean. Cleanliness achieves two objectives. It ensures that
the working fluid will wet the materials, and that no foreign matter is
present which could hinder capillary action or create incompatibilities.

The cleaning procedure depends upon the material used, the process undergone
in manufacturing and locating the wick, and the requirements of the working
fluid, some of which wet more readily than others. In the case of wick/wall
assemblies produced by processes such as sintering or diffusion bonding,
carried out under an inert gas or vacuum, the components are cleaned during
the bonding process, and provided that the time between this process and
final assembly is short, no further cleaning may be necessary.

If the working fluid is a solvent, such as acetone, no extreme precautions
are necessary to ensure good wetting, and an acid pickle followed by a rinse
in the working fluid appears to be satisfactory. However, cleaning
procedures become more rigorous as one moves up the operating temperature
range to incorporate liquid metals as working fluids.

The pickling process for stainless steel involves immersing the components in
a solution of 50% nitric acid and 5% hydrofluoric acid. This is followed by
a rinse in demineralised water. If the units are to be used in conjunction
with water, the wick should then be placed in an electric furnace and heated
in air to $400^{\circ}C$ for one hour. At this temperature grease is either volatised
or decomposed, and the resulting carbon burnt off to form carbon dioxide.
Since an oxide coating is required on the stainless steel, it is not
necessary to use an inert gas blanket in the furnace.

Nickel may undergo a similar process to that described above for stainless
steel but pickling should be carried out in a 25% nitric acid solution.
Pickling of copper demands a 50% phosphoric acid 50% nitric acid mixture.

Cleanliness is difficult to quantify, and the best test is to add a drop of
demineralised water to the cleaned surface. If the drop immediately spreads
across the surface, or is completely absorbed into the wick, good wetting has

occurred, and satisfactory cleanliness has been achieved.

Stainless steel wicks in long heat pipes sometimes create problems in that furnaces of sufficient size to contain the complete wick may not be readily available.   In this case a flame cleaning procedure may be used, whereby the wick is passed through a Bunsen flame as it is fed into the container.

An ultrasonic cleaning bath is a useful addition for speeding up the cleaning process, but is by no means essential for low temperature heat pipes.   As with this process or any other associated with immersion of the components in a liquid to remove contaminants, the debris will float to the top of the bath and must be skimmed off before removing the parts being treated.   If this is not done, the parts could be recontaminated as they are removed through this layer.   Electro-polishing may also be used to aid cleaning of metallic components.

Ceramic wick materials are generally exceptionally clean when received from the manufacturer, owing to the production process used to form them, and therefore need no treatment provided that handling during assembly of the heat pipe is under clean conditions.

It is important, particularly when water is used as the working fluid, to avoid skin contact with the heat pipe components.   Slight grease contamination can prevent wetting, and the use of surgical gloves for handling is advisable.   Wetting can be aided by additives (wetting agents) (4.1) applied to the working fluid, but this can introduce compatibility problems and also affect surface tension.

4.1.4 Material outgassing.

When the wick or wall material is under vacuum, gases will be drawn out, particularly if the components are metallic.   If not removed prior to sealing of the heat pipe, these gases could collect in the heat pipe vapour space.   The process is known as outgassing.

While outgassing does not appear to be a problem in low temperature heat pipes for applications which are not too arduous, high temperature units ($>400^{\circ}$C) and pipes for space use should be outgassed in the laboratory prior to filling with working fluid and sealing.

The outgassing rate is strongly dependent on temperature, increasing rapidly as the component temperature is raised. It is advisable to outgas components following cleaning, under vacuum at a baking temperature of about $400^{\circ}C$. Following baking the system should be vented with dry nitrogen. The rate of outgassing depends on the heat pipe operating vapour pressure, and if this is high the outgassing rate will be restricted.

If the heat pipe has been partially assembled prior to outgassing, and the end caps fitted, it is necessary to make sure that no welds etc. leak, as these could produce misleading results as to outgassing rate. It will generally be found that analysis of gases escaping through a leak will show a very large air content, whereas those brought out by outgassing will contain a substantial water vapour content. A mass spectrometer can be used to analyse these gases. Leak detection is covered in Section 4.1.6 below.

The outgassing characteristics of metals can differ considerably. The removal of hydrogen from stainless steel, for example, is much easier to effect than its removal from aluminium. Aluminium is particularly difficult to outgas, and can hold comparatively large quantities of non-condensables. In one test it was found (4.2) that gas was suddenly released from the aluminium when it approached red heat under vacuum. 200 grams of metal gave 89.5 cc of gas at NTP, 88 cc being hydrogen and the remainder carbon dioxide. It is also believed that aluminium surfaces can retain water vapour even when heated to $500^{\circ}C$ or dried over phosphorus pentoxide. This could be particularly significant because of the known incompatibility of water with aluminium. (See Section 4.1.12 for high temperature heat pipes).

4.1.5 Fitting of wick and end caps.

Cleaning of the heat pipe components is best carried out before insertion of the wick, as it is then easy to test the wick for wettability. Outgassing may be implemented before assembly or while the heat pipe is on the filling rig (see Section 4.1.8).

In cases where the wick is an integral part of the heat pipe wall, as in the case of grooves, sintered powders, or diffusion bonded meshes, cleaning of the heat pipe by flushing through with the appropriate liquid is convenient, prior to the welding of the end caps.

If a mesh wick is used, and the mesh layers are not bonded to one-another or
to the heat pipe wall, particularly when only a fine mesh is used, a coiled
spring must be inserted to retain the wick against the wall. This is readily
done by coiling the spring tightly around a mandrel giving a good internal
clearance in the heat pipe. The mandrel is inserted into the pipe, the spring
tension released, and the mandrel then removed. The spring will now be
holding the wick against the wall. Typically the spring pitch is about 1 cm.
(In instances where two mesh sizes may be used in the heat pipe, say 2 layers
of 400 mesh and 1 layer of 100 mesh, the liquid vapour interface must always
be in the 400 mesh to achieve maximum capillary rise. It is therefore
advisable to wrap the 400 mesh over the end of the 100 mesh, as shown in
Fig. 4.6).

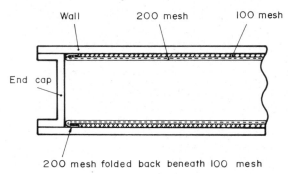

Fig. 4.6 Sealing of mesh at end of heat pipe

The fitting of end caps is normally carried out by argon arc welding. This
need not be done in a glove box and is applicable to copper, stainless steel
and aluminium heat pipes. The advantage of welding over brazing or soldering
is that no flux is required, therefore the inside of cleaned pipes do not
suffer from possible contamination. However, possible inadequacies of the
argon shield, in conjunction with the high temperatures involved can lead to
local material oxidisation which may be difficult to remove from the heat
pipe interior. Assembly in a glove box filled with argon would overcome this
but would be expensive. The use of a thermal absorbent paste such as Rocol HS
to surround the area of heat pipe local to the weld can considerably reduce
the amount of oxide formed.

Electron beam welding may also be used for heat pipe assembly, but this added
expense cannot be justified in most applications.

4.1.6 **Leak detection.**   All welds on heat pipes should be checked for leaks.
If quality control is to be maintained, a rigorous leak check procedure is
necessary because a small leak which may not affect heat pipe performance
initially could make itself felt over a period of months.

The best way to test a heat pipe for leaks is to use a mass spectrometer,
which can be used to evacuate the heat pipe to a very high vacuum, better
than 10^{-5} torr, using a diffusion pump. The weld area is then tested by
directing a small jet of helium gas onto it. If a leak is present, the gauge
head on the mass spectrometer will sense the presence of helium once it
enters the heat pipe. After an investigation of the weld areas and location
of the general leak area(s), if present, a hypodermic needle can then be
attached to the helium line and careful traversing of the suspected region
can lead to very accurate identification of the leak position, possibly
necessitating only a very local rewelding procedure to seal it.

Obviously if a very large leak is present, the pump on the mass spectrometer
may not even manage to obtain a vacuum better than 10^{-2} or 10^{-3} torr.
Porosity in weld regions can create conditions leading to this, and may point
to impure argon, or an unsuitable welding filler rod.

It is possible, if the leak is very small, for water vapour from the breath
to condense and block, albeit temporarily, the leak. It is therefore
important to keep the pipe dry during leak detection.

4.1.7 **Preparation of the working fluid.**   It is necessary to treat the
working fluid used in a heat pipe with the same care as that given to the
wick and container.

The working fluid should be the most highly pure available, and further
purification may be necessary following purchase. This may be carried out
by distillation. In the case of low temperature working fluids such as
acetone, methanol, and ammonia in the presence of water can lead to incompatibilities, and the minimum possible water content should be achieved.

Some brief quotations from a treatise on organic solvents (4.3) highlight the
problems associated with acetone and its water content: "Acetone is much

more reactive than is generally supposed. Such mildly basic materials as
alumina gel cause aldol condensation to 4-hydroxy-4-methyl-2-pentanone,
(diacetone alcohol), and an appreciable quantity is formed in a short time if
the acetone is warm. Small amounts of acidic material, even as mild as
anhydrous magnesium sulphate, cause acetone to condense".

"Silica gel and alumina increased the water content of the acetone, presum-
ably through the aldol condensation and subsequent dehydration. The water
content of acetone was increased from 0.24 to 0.46% by one pass over alumina.
All other drying agents tried .... caused some condensation".

Ammonia has a very great affinity for water, and it has been found that a
water content of the order of <10 ppm is necessary to obtain satisfactory
performance. ICI are able to supply high purity ammonia, but exposure to air
during heat pipe filling must be avoided.

The above examples are extreme, but serve to illustrate problems which can
arise when handling procedures are relaxed.

A procedure which is recommended for all heat pipe working fluids used up to
200°C is freeze-degassing. This process removes all dissolved gases from the
working fluid, and if the gases are not removed they could be released
during heat pipe operation and collect in the condenser section. Freeze-
degassing may be carried out on the heat pipe filling rig described in
Section 4.1.8, and is a simple process. The fluid is placed in a container
in the rig directly connected to the vacuum system and is frozen by surrounding the container with a flask containing liquid nitrogen. When the working
fluid is completely frozen the container is evacuated and resealed and the
liquid nitrogen flask removed. The working fluid is then allowed to thaw and
dissolved gases will be seen to bubble out of the liquid. The working fluid
is then refrozen and the process repeated. All gases will be removed after
three or four freezing cycles.

The liquid will now be in a sufficiently pure state for insertion into the
heat pipe.

4.1.8 Heat pipe filling.   A flow diagram for a rig which may be used for
heat pipe filling is shown in Fig. 4.7.   The rig may also be used to carry
out the following processes:

> Working fluid degassing
> Working fluid metering
> Heat pipe degassing
> Heat pipe filling with inert gas.

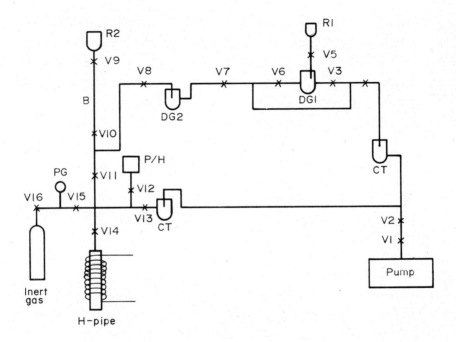

Fig. 4.7   A heat pipe filling rig layout

Before describing the rig and its operation, it is worth mentioning the
general requirements when designing vacuum rigs.   The material of construction for pipework is generally either glass or stainless steel.   Glass has
advantages when handling liquids in that the presence of liquid droplets in
the ductwork can be observed and their vaporisation under vacuum noted.
Stainless steel has obvious strength benefits, and must be used for all high
temperature work, together with high temperature packless valves such as Hoke
bellows valves.   The rig described below is for low temperature heat pipe
manufacture.

Valves used in vacuum rigs should preferably have '0' ring seals, and it is
important to ensure that the ductwork is not too long or has a small diameter, as this can greatly increase evacuation times.

The vacuum pump may be the diffusion type or a sorption pump containing a
molecular seive which can produce vacuums as low as 10^{-4} torr. It is, of
course, advisable to refer to experts in the field of high vacuum technology
when considering designing a filling rig.

Description of rig: The heat pipe filling rig described below is made using
glass for most of the pipework. Commencing from the right hand side, the
pump is of the sorption type, which is surrounded by a polystyrene container
of liquid nitrogen when a vacuum is desired. Two valves are fitted above
the pump, the lower one being used to disconnect the pump when it becomes
saturated. (The pump may be cleaned by baking out in a furnace for a few
hours). Above valve V2 a glass-to-metal seal is located and the rest of the
pipework is glass. Two limbs lead from this point, both interrupted by cold
traps, in the form of small glass flasks, which are used to trap stray liquid
and any impurities which could affect other parts of the rig or contaminate
the pump. The cold traps are formed by surrounding each flask with a
container of liquid nitrogen.

The upper limb includes provision for adding working fluid to the rig and
two flasks are included (DG1 and DG2) for degassing the fluid. The section
of the rig used for adding fluid can be isolated once a sufficient quantity
of fluid has been passed to flask DG2 and thence to the burette between
valves V9 and V10.

The lower limb incorporates a Pirani head which is used to measure the degree
of vacuum in the rig. The heat pipe to be filled is fitted below the burette,
and provision is also made to electrically heat the pipe to enable outgassing
of the unit to be carried out on the rig (see also Section 4.1.4). An
optional connection can be made via valve V15 to permit the loading of inert
gas into the heat pipe for variable conductance types.

Procedure for filling a heat pipe: The following procedure may be followed
using this rig for filling, for example, a copper/ethanol heat pipe.

(i)   Close all valves linking rig to atmosphere (V5, V9, V14,
        V15).

(ii)   Attach sorption pump to rig via valves V1 and V2, both
        of which should be closed.

(iii)   Surround the pump with liquid nitrogen, and also top up
        the liquid nitrogen containers around the cold traps.
        (It will be found that the LN_2 evaporates quickly initially,
        and regular topping-up of the pump and traps will be
        necessary).

(iv)   After approximately 30 minutes, open valves V1 and V2,
        commencing rig evacuation. Evacuate to about 0.010 mm Hg,
        the time to achieve this depending on the pump capacity,
        rig cleanliness and rig volume.

(v)   Close valves V4 and V6, and top up reservoir R1 with
        ethanol.

(vi)   Slowly crack valve V5 to allow ethanol into flask DG1.
        Reclose V5 and freeze the ethanol using a flask of LN_2
        around DG1.

(vii)   When all the ethanol is frozen, open V4 and evacuate.
        Close V4 and allow ethanol to melt. Any gases will bubble
        out of the ethanol as it melts. The ethanol is then re-
        frozen.

(viii) Open V4 to remove gas.

(ix)   Close V4, V3 and V8; open V6 and V7. Place LN_2 container
        around flask DG2.

(x)   Melt the ethanol in DG1 and drive it into DG2. (This is
        best carried out by carefully heating the frozen mass using
        a hair dryer. Warming of the ductwork between DG1 and DG2
        and up to V4, will assist).

(xi)   The degassing process may be repeated in DG2 until no more
        bubbles are released. V4 and V6 are now closed, isolating
        DG1.

(xii)   Close V7 and V11; open V8 and V10, and drive the ethanol

into the burette as in (x). Close V10 and V8 and open V11.
The lower limb and upper limb back to V8 are now brought to a
to a high vacuum (≈ 0.005 mm Hg).

The heat pipe to be filled should now be attached to the rig. In cases where
the heat pipe does not have its own valve, the filling tube may be connected
to the rig below V14 using thick walled rubber tubing, or in cases where this
may be attacked by the working fluid, another flexible tube material or a
metal compression or 'O' ring coupling. If a soft tube material is used, the
joints should be covered with a silicone-based vacuum grease to ensure no
leaks.

The heat pipe may be evacuated by opening valve V14. Following evacuation
which should take only a few minutes, depending on the diameter of the
filling tube, the heat pipe may be outgassed by heating. This can be done by
surrounding the heat pipe with electric heating tape, and applying heat until
the Pirani gauge returns to the maximum vacuum obtained before heating
commenced. (It is worth emphasising the fact that, depending on the diameter
of the heat pipe and filling tube, the pressure recorded by the Pirani is
likely to be less than that in the heat pipe. It is preferable from this
point of view to have a large diameter filling tube).

To prepare the heat pipe for filling, the lower end is immersed in liquid
nitrogen so that the working fluid, which flows towards the coldest region,
will readily flow to the heat pipe base. Valve V10 is then cracked and the
correct fluid inventory (in most cases enough to saturate the wick plus a
small excess) allowed to flow down into the heat pipe. Should fluid stray
into valve seats or other parts of the rig, local heating of these areas
using the hot air blower will evaporate any liquid, which should then
condense and freeze in the heat pipe. A further freeze-degassing process may
be carried out with the fluid in the heat pipe, allowing it to thaw with V14
closed, re-freezing and then opening V14 to evacuate any gas. The heat pipe
may then be sealed.

4.1.9 Heat pipe sealing.   Unless the heat pipe is to be used as a demonstration unit, or for life testing, in which cases a valve may be retained on
one end, the filling tube must be permanently sealed.

With copper this is conveniently carried out using a tool which will crimp
and cold weld the filling tube. A typical crimp obtained with this type of
tool is shown in Fig. 4.8, and the force to operate this is applied manually.

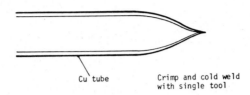

Cu tube   Crimp and cold weld
                     with single tool

Fig. 4.8   Crimped and cold welded seal

If stainless steel or aluminium are used as the heat pipe filling tube
material, crimping followed by argon arc welding is a more satisfactory
technique. Once the desired vacuum has been attained and the fluid injected,
two 0.5 in. jaws are brought into contact with the evacuating tube and the
latter is flattened. The heat pipe is then placed between two 0.25 in. thick
jaws located at the lower half of the 0.5 in. flattened section. Sufficient
load is placed on the evacuating tube to temporarily form a vacuum-tight seal
and the remaining 0.25 in. flattened section is simultaneously cut through
and welded using an argon arc torch. The 0.25 in crimping tool, which fits
between the jaws of a standard vice, is shown in Fig. 4.9. Results obtained
are as shown in Fig. 4.10.

Fig. 4.9   Jaws for crimping prior to welding

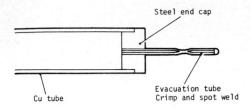

Fig. 4.10 Crimped and argon-arc welded end

Following sealing the filling tube may be protected by a cap having an outer
diameter the same as that of the heat pipe wall. The cap may be filled with
solder, a metal-loaded resin, or any other suitable material.

4.1.10 Summary of assembly procedures.   The following is a list of the
procedures described above which should be followed during heat pipe assembly.

    (i)   Select container material
    (ii)   Select wick material and form
    (iii) Fabricate wick and end caps etc.
    (iv)   Clean wick, container and end caps
    (v)   Outgas metal components
    (vi)   Insert wick and locate
    (vii) Weld end caps
    (viii) Leak check welds
    (ix)   Select working fluid
    (x)   Purify working fluid (if necessary)
    (xi)   De-gas working fluid
    (xii) Evacuate and fill heat pipe
    (xiii) Seal heat pipe

It may be convenient to weld the blank end cap before wick insertion, and in
cases of sintered and diffusion bonded wicks the outgassing may be done with
the wick in place in the container.

4.1.11 Heat pipes containing inert gas.   Heat pipes of the variable conductance type (see Chapter 6) contain an inert gas in addition to the normal
working fluid, and an additional step in the filling process must be carried

out. The additional features on the filling rig to cater for inert gas
metering are shown in Fig. 4.7.

The working fluid is inserted into the heat pipe in the normal way, and then
the system is isolated and the line connecting the heat pipe to the inert gas
bottle is opened and the inert gas bled into the heat pipe. The pressure
increases as the inert gas quantity in the heat pipe is raised, as indicated
by the pressure gauge in the gas line. The pressure appropriate to the
correct gas inventory may be calculated, taking into account the partial
pressure of the working fluid vapour in the heat pipe (see Chapter 6 for mass
calculation) and when this is reached the heat pipe is sealed in the normal
manner.

4.1.12 Liquid metal heat pipes.

The early work on heat pipes was concerned
with the application to thermionic generators and is described in Chapter 7.
For this application there are two temperature ranges of interest, the
emitter range of 1400 - 2000°C and the collector range of 500 - 900°C. In
both temperature ranges liquid metal working fluids are required and there is
now a considerable body of information on the fabrication and performance of
such heat pipes. More recently heat pipes operating in the lower temperature
range have been used to transport heat from the heater to the multiple cylinders of a Stirling engine and for industrial ovens. A large range of material
combinations have been found suitable in this temperature range and
compatibility and other problems are well understood. The alkali metals are
used with containment materials such as stainless steel, nickel, niobium-
zirconium alloys and other refractory metals. Lifetimes of greater than
20 000 hours are reported (4.4). Grover (4.5) reports on the use of a light
weight pipe made from beryllium and using potassium as the working fluid.
The beryllium was inserted between the wick and wall of a pipe both made from
niobium-1% zirconium. The pipe operated at 750°C for 1200 hours with no
signs of attack, alloying or mass transport.

The construction of heat pipes for use in the higher temperature range has
proved more difficult, some success has been achieved and is discussed below.

Very high performance, long life, liquid metal heat pipes can be constructed
with some confidence; they are however expensive. Hence before commencing

the design of a liquid metal heat pipe, it is important to decide what is to
be required from it. It frequently happens that an application does not
require the pipe to pump against a gravity head so that a thermal syphon will
be adequate. This greatly reduces the importance of working fluid purity.
Again short operating life at low rating will enable cheaper and less time
consuming fabrication methods to be adopted. If gas buffering is possible a
simpler crimped seal arrangement can be used.

Liquid metal heat pipes for the temperature range 500 - 1100°C: In this temperature range potassium and sodium are the most suitable working fluids and
stainless steel is selected for the container. Fig. 4.11 shows a typical
sodium in stainless steel heat pipe operating at Reading University and
described by Rice and Jennings (4.6). The construction and fabrication of
this pipe will be described to indicate the processes involved. The heat
pipe container was made from type 321 (EN58B) stainless steel tube 2.5 cm
diameter and 0.9 mm wall thickness. The capillary structure was two layers
of 100 mesh stainless steel having a wire diameter of 0.1016 mm and an
aperture size of 0.152 mm. The pipe was 0.9 m in length and the wick welded
by spot welds using a special tool built for the purpose.

Fig. 4.11   Sodium heat pipe operating at Reading University (Courtesy Reading University).

Cleaning and filling: The following cleaning process was followed.

(i)   Wash with water and detergent

(ii)   Rinse with demineralised water

(iii)   Soak for 30 minutes in 1:1 mixture of hydrochloric acid
         and water

(iv)   Rinse with demineralised water

(v)   Soak for 20 minutes in an ultrasonic bath filled with
         acetone and repeat with a clean fluid.

After completion of the welds and brazes this procedure was repeated. Argon
arc welding was used throughout, and after leak testing the pipe was outgassed at a temperature of 900°C and a pressure of 10^{-5} torr for a period of
several hours in order to remove gases and vapours.

Various methods may be used to fill the pipe with liquid metal including

(i)   Distillation, sometimes from a getter sponge to remove
         oxygen

(ii)   Breaking an ampoule contained in the filler pipe by
         distortion of the filler pipe

Distillation is essential if a long life is required. The method adopted for
the pipe being described was as follows:

(iii)   99.9% industrial sodium was placed in a glass filter tube
         attached to the filling tube of the heat pipe. A by-pass
         to the filter allowed the pipe to be initially evacuated
         and outgassed. The filling pipe and heat pipe were
         immersed in the heated liquid paraffin bath to raise the
         sodium above its melting point. The arrangement is shown
         in Fig. 4.12.

Finally the by-pass valve is closed and a pressure applied by means of helium
gas to force the molten sodium through the filter and into the heat pipe.

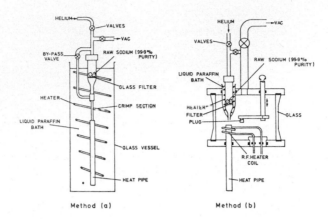

Fig. 4.12   Liquid metal heat pipe filling (Courtesy
Reading University)

Sealing: For liquid metal heat pipes at Reading University, the technique
of plug sealing has been adopted, as shown in Fig. 4.13. A special rig has
been constructed which allows for outgassing of an open ended tube and sodium
filling by the filtering method described above. On completion of the
filling process the end sealing plug, supported by a swivel arm within the
filling chamber, is swung into position and placed within the heat pipe. The
plug is then induction heated to effect a brazed vacuum seal. The apparatus
and sequence of operation is illustrated in Fig. 4.12. The end sealing plug
is finally argon arc welded - after removal of the heat pipe from the filling
apparatus.

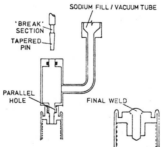

Fig. 4.13   Plug sealing technique for sealing liquid
metal heat pipes (Courtesy Reading University)

Operation: It has been found that wetting of the wick structure does not occur immediately and it was necessary to heat the pipe as a thermal syphon for several hours at 650°C. Heating was by an R.F. induction heater over a length of 10 cm. Temperature profiles are given in Fig. 4.14 for heat inputs of 1.2 kW and 1.4 kW. Before sealing the heat pipe was filled with helium at a pressure of 20 torr to protect the copper crimp by the resulting gas buffer. It is seen that the gas buffer length is approximately proportional to the power input as might be expected.

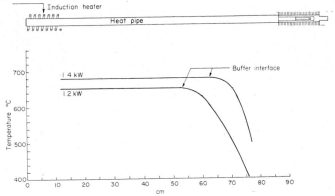

Fig. 4.14   Temperature profiles along a sodium heat pipe (Courtesy Reading University)

The start up of the heat pipe after conditioning was interesting. In the thermal syphon mode, that is with the pipe vertical and heated at the bottom, there were violent temperature variations associated with boiling in the evaporator zone. This was not experienced when the heater was at the top of the heat pipe.

Similar work has been reported by other authors. An interesting method for making rigid thin walled wicked pipes is described by Vinz et al (4.7). Previous work on mesh wicks has included methods such as spot welding, drawing and drawing and sintering. The first method does not give uniform adhesion and drawing methods cannot be used for very fine wicks (<200 - 400 mesh) because of damage. Vinz's method consists of winding a screen strip spirally on a mandrel and sintering it under simultaneous axial pulling and twisting. Gauze of 508 x 3600 mesh has been used successfully to give pore diameters of 10μ reproducible to $\pm$ 10% and with a free surface for evaporation of 15 - 20%.

Broached grooves can be used either alone or with gauze wicks.

Very high temperature liquid metal heat pipes >1200°C: At the lower end of
the range lithium is preferred as the working fluid and niobium-zirconium or
tantalum as the container material. At higher temperatures silver may be used
as the working fluid with tungsten or rhenium as the container material. Data
on the compatibility and lifetime of heat pipes made from these materials is
given in Chapter 3. Such refractory materials have a high affinity for oxygen
and must be operated in a vacuum or inert gas.

Busse and his collaborators have carried out a considerable programme on
lithium and silver working fluid heat pipes, and the techniques used for
cleaning, filling, fabrication and sealing are described in Refs. 4.8 and 4.9.

Gettering: Oxides can be troublesome in liquid metal heat pipes since they
will be deposited in the evaporator area. Dissolved oxygen is a particular
problem in lithium heat pipes since it causes corrosion of the container
material. Oxygen can arise both as an impurity in the heat pipe fluid and
also from the container and wick material. A number of authors report the
use of getters. For example Busse (4.9) uses a zirconium sponge from which
he distills his lithium into the pipe. Calcium can also be used for
gettering.

4.1.13 Safety aspects.

Whilst there are no special hazards associated with
heat pipe construction and operation there are a number of aspects which
should be borne in mind.

Where liquid metals are employed standard handling procedures should be
adopted. The affinity of alkali metals for water can give rise to problems;
a fire was started in one laboratory when a sodium in stainless steel pipe
distorted releasing the sodium and at the same time fracturing a water pipe.

Mercury is a highly toxic material and its saturated vapour density at atmospheric pressure is many times the recommended maximum tolerance.

One danger which is sometimes overlooked is the high pressure which may occur
in a heat pipe when it is accidentally raised to a higher temperature than

its design value. Water is particularly dangerous in this respect. The
critical pressure of water is 220 bar and occurs at a temperature of 374°C.
When a water in copper heat pipe sealed by a soldered plug was inadvertently
overheated, both the 30 cm long heat pipe and the plug were ejected from the
clamps at very high velocity and could well have had fatal results. It is
imperative that a release mechanism such as a crimp seal be incorporated in
such heat pipes.

Cryogenic heat pipes employing fluids such as liquid air should have special
provision for pressure release or be of sufficient strength since they are
frequently allowed to rise to room temperature when not in use. The critical
pressure of nitrogen is 34 bar.

4.2 Heat Pipe Life Test Procedures

Life testing and performance measurements on heat pipes, in particular when
accelerated testing is required, are about the most important factors in their
selection.

Life tests on heat pipes are commonly regarded as being primarily concerned
with the identification of any incompatibilities which may occur between the
working fluid and wick and wall materials.   The ultimate life test, however,
would be in the form of a long term performance test under conditions
appropriate to those in the particular application. If this is done it is
difficult however, in cases where the wick is pumping against gravity, to
accelerate the life test by increasing, say the evaporator heat flux, as this
could well cause heat pipe failure owing to the fact that it is likely to be
operating well in excess of its design capabilities. This, therefore, necessitates operation in the reflux mode.

There are many factors to be taken into account when setting up a full life
test programme, and the relative merits of the alternative techniques available are discussed below.

4.2.1 Variables to be taken into account during life tests.   The number of
variables to be considered when examining the procedure for life tests on a
particular working fluid/wick/wall combination is very extensive, and would
require a large number of heat pipes to be fully comprehensive.

Several of these may be discounted because of existing available data on
particular aspects, but one important point which must be emphasised is the
fact that quality control and assembly techniques inevitably vary from one
laboratory to another, and these differences can be manifest in differing
compatibility data and performance.

The factors to be taken into account when investigating the life of a heat
pipe are shown in Fig. 4.15. (The non-condensable gas is included to cover
variable conductance heat pipes).

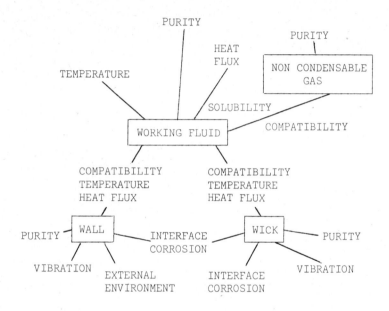

Fig. 4.15   Heat pipe life test factors

The working fluid: The selection of the working fluid must take into account
the following factors which can all be investigated by experiments:

(i)   Purity - the working fluid must be free of dissolved gases
        and other liquids, for example water. Such techniques as
        freeze-degassing and distillation are available to purify
        the working fluid. It is important to ensure that handling
        of the working fluid following purification does not expose
        it to contaminants.

(ii)   Temperature - some working fluids are sensitive to opera-
        ting temperature. If such behaviour is suspected, the
        safe temperature band must be identified.

(iii)   Heat Flux - high heat fluxes can create vigorous boiling
        action in the wick which can lead to errosion.

(iv)   Compatibility with wall and wick - the working fluid must
        not react with the wall and wick. This can also be a
        function of temperature and heat flux, the tendency for
        reactions to occur generally increasing with increasing
        temperature or flux.

(v)   Non-condensable gas - in the case of variable conductance
        heat pipes, where a non-condensable gas is used in con-
        junction with the working fluid, the selection of the two
        fluids must be based on compatibility and also on the
        solubility of the gas in the working fluid. (In general
        this data is available from the literature, but in specific
        arterial design the effect of solubility may only be
        apparent after experimentation).

The heat pipe wall: In addition to the interface with the heat pipe working
fluid, as discussed above, the wall and associated components such as end
caps have their own particular requirements with regard to life, and also
interface with the wick. The successful operation of the heat pipe must take
into account the following:

(i)   Vibration and acceleration - the structure must be able to
        to withstand any likely vibrations and accelerations and
        any qualification procedures designed to ensure that the
        units meet these specifications should be regarded as an
        integral part of any life test programme.

(ii)   Quality assurance - the selection of the outer case material
        should be based on the purity, or at least the known alloy
        specification of the metal used.

(iii)   External environment - the external environment could
        affect the case material properties, or cause degradation
        of the outer surface. This should also be the subject of

HP—G

life test investigation if any deterioration is suspected.

(iv)   Interface corrosion - it is possible that some corrosion
        could occur at metallic interfaces, particularly where
        dissimilar metals are used, in the presence of the working
        fluid.

The wick: The heat pipe wick is subjected to the same potential hazards as
the heat pipe wall, with the exception of external attack. Vibration is much
more critical however, and the wick itself contains, in most cases, many
interfaces where corrosion could occur.

4.2.2 Life test procedures.   There are many ways of carrying out life tests,
all having the same aim, namely to demonstrate that the heat pipe can be
expected to last for its design life with an excellent degree of certainty.

The most difficult part of any life test programme is the interpretation of
the results and the extrapolation of these results to predict long term
performance. (One technique used for extrapolating results obtained from gas
generation measurements is described in Section 4.2.3).

The main disadvantage of carrying out life tests of one particular combination
of materials, be the test accelerated or at design load, is the fact that if
any reaction does occur, insufficient data is probably available to enable
one to explain the main causes of the degradation. For example, in some life
tests carried out at IRD, diacetone alcohol was formed as a result of acetone
degradation. It was not possible without further testing over a considerable
period, however, to state whether this phenomenon was a function of operating
temperature, as life tests on identical units operating at several different
vapour temperatures will have to be carried out. It is even possible that
comprehensive life test programmes may never provide the complete answer to
some questions, new aspects being found during each study.

Effect of heat flux.   The effect of heat flux on heat pipe lives and performance can only really be investigated using units in the reflux mode, where
fluxes well in excess of design values may be applied.

By setting up experiments involving a number of heat pipes operating at the same vapour temperature but with differing evaporator heat fluxes, one can later examine the inner surface of the evaporator for corrosion etc.

If carried out in a representative heat pipe, performance tests could be carried out at regular intervals during the life tests.

Effect of temperature: Compatibility and working fluid make-up can both be affected by the operating temperature of the heat pipe. It is therefore important to be able to discriminate between any effects resulting from temperature levels.

As discussed in Section 4.2.3 the effect of temperature on non-condensable gas generation can, under certain circumstances, be predicted over a long period as a result of much shorter tests on the representative heat pipes.

Compatibility: As opposed to the effect of heat flux or temperature on the working fluid alone, it is necessary to investigate the compatibility of the working fluid with the wall and wick materials.

Here one is looking for reactions between the materials which could change the surface structure in the heat pipe, generate non-condensable gas, or produce impurities in the form of deposits which could affect evaporator performance. Of course all three phenomena could occur at the same time, to differing degrees, and this can make the analysis of the degradation much more complex.

Compatibility tests can be carried out at design conditions on a heat pipe operating horizontally or under a tilt against gravity. To be meaningful such tests should continue over a period of years, but if compatibility is shown to be satisfactory over, say, a three year period, some conclusions can be made concerning likely behaviour over a much longer life. Accelerated compatibility tests could also be performed, with occasional tests in the heat pipe mode to check on the design performance.

Other factors: The life of a heat pipe can be affected by assembly and cleaning procedures, and it is important to ensure that life test pipes are fully

representative as far as assembly techniques are concerned. The working
fluid used must, of course, be of the highest purity.

Another feature of life testing is the desirability of incorporating valves
on the pipes so that samples of gas etc. can be taken out without necessarily
causing the unit to cease functioning. One disadvantage of valves is the
introduction of a possibly new incompatibility, that of the working fluid
and valve material, although this can be ruled out with modern stainless
steel valves.

When testing in the heat pipe mode, a valve body can become full of working
fluid, which may be difficult to remove. This should be taken into account
when carrying out such tests, in case depletion of the wick or artery system
occurs.

4.2.3 Prediction of long term performance from accelerated life tests.

One of the major drawbacks of accelerated life tests has been the uncertainty
associated with the extrapolation of the results to estimate performance over
a considerably longer period of time. Baker (4.10) has correlated data on
the generation of hydrogen in stainless steel heat pipes, using an Arrhenius
plot, with some success, and this has been used to predict non-condensable gas
generation over a 20 year period.

The data was based on life tests carried out at different vapour temperatures
over a period of 2 years, the mass of hydrogen generated being periodically
measured. Vapour temperatures of 100, 200 and 300°F were used, 5 heat pipes
being tested at each temperature.

Baker applied Arrhenius plot to these results, which were obtained at the Jet
Propulsion Laboratory, in the following way.

The Arrhenius model is applicable to activation processes, including
corrosion, oxidation, creep and diffusion. Where the Arrhenius plot is valid,
the plot of the log of the response parameter (F) against the reciprocal of
absolute temperature is a straight line.

The response parameter is defined by the equation:

$$F = \text{Const.} \times \exp. -A/kT \qquad \qquad \ldots 4.1$$

where

A = the reaction activation energy

k = the Boltzmann constant $(1.38 \times 10^{-23} \text{ J/K})$

T = the absolute temperature.

For the case of the heat pipe, Baker described the gas generation process as:

$$\dot{m}(t,T) = f(t)\, F(T) \qquad \qquad \ldots 4.2$$

where

$\dot{m}$ is the mass generation rate

t   denotes time

F(T) is given in eqn 4.1

By plotting the mass of hydrogen generated in each heat pipe against time, with results at different temperatures, one can use these figures to obtain a universal curve, presenting the mass of hydrogen generated as a function of time × shift factor, which will be a straight line on logarithmic paper. Finally the shift factors are plotted against the reciprocal of absolute temperature for each temperature examined, and the slope of this curve gives the activation energy A in equation 4.1.

The mass of hydrogen generated at any particular operating temperature can then be determined by using the appropriate value of shift factor. Baker concluded that stainless steel/water heat pipes could operate for many years at temperatures of the order of 60°F, but at 200°F the gas generation would be excessive.

It is probable that this model could be applied to other wall/wick/working fluid combinations, the only drawback being the large number of test units needed for accurate predictions. The minimum is of the order of 12, results being obtained at 3 vapour temperatures, 4 heat pipes being tested at each temperature.

A recent study was concerned with the evolution of hydrogen in nickel/water heat pipes. Anderson used a corrosion model to enable him to predict the behaviour of heat pipes over extended periods, based on accelerated life tests, following Baker's method. (4.11).

He argued that oxidation theory predicts that passivating film growth occurs with a parabolic time dependence and an exponential temperature dependence.

Anderson gives the following values for A, the reaction activation energy:

       Stainless steel (304)/water 8.29×10^{-20} J
       Nickel/water 10.3×10^{-20} J

and confirms Baker's model.

4.2.4 A life test programme. A life test programme must provide detailed data on the effects of temperature, heat flux, and assembly techniques on the working fluid, and the working fluid/wall and wick material compatibility.

The alternative techniques for testing have been discussed in Section 4.2.2 and it now remains to formulate a programme which will enable sufficient data to be accumulated to enable the life of a particular design of heat pipe to be predicted accurately.

Each procedure may be given a degree of priority (numbered 1 to 3, in decreasing order of importance) and these are presented in Table 4.1 together with the MINIMUM number of units required for each test.

The table is self-explanatory.

This programme should provide sufficient data to enable one to confidently predict long term performance of a heat pipe, based on an Arrhenius plot, and the maximum allowable operating temperature, based on fluid stability. The programme involves a considerable amount of testing, but the cost should be weighed against satisfactory heat pipe performance over its life in applications where reliability is of prime importance.

TABLE 4.1   HEAT PIPE LIFE TEST PRIORITIES

Priority	Minimum No units to be tested	Number with valves	Test specification
1	-	-	Cleanliness of materials
1	-	-	Purity of working fluid
1	-	-	Sealing of case
1	-	-	Outgassing
2	2	2	Refluxing - vapour temperature at maximum design
1	4 at each temp.	all	Refluxing - temperature range up to maximum design (to include bonding temperatures)
3	2	2	Refluxing - heat flux at maximum design
2	2	1	Heat pipe mode - intermittent tests between refluxing
1	2	1	Heat pipe mode - long term continuous performance test
1*	2	0	Heat pipe mode - vibration test with intermittent performance tests
1	2	2	(VCHP) - solubility of gas in working fluid and effect on artery

* where applicable

4.3 Heat Pipe Performance Measurements. (See also Section 4.1.12)

The measurement of the performance of heat pipes is comparatively easy and requires in general equipment available in any laboratory engaged in heat transfer work.

Measurements are necessary to show that the heat pipe meets the requirements laid down during design. The limitations to heat transport, described in Chapter 2, and presented in the form of a performance envelope, can be investigated, as can the degree of isothermalisation. A considerable number of variables can be investigated by bench testing, including orientation with respect to gravity, vapour temperature, evaporator heat flux, start-up, vibrations and accelerations.

4.3.1 The test rig .   A typical test rig is shown diagrammatically in
Fig. 4.16.   The rig has the following features and facilities:

    (i)   Heater for evaporator section

    (ii)   Wattmeter for power input measurement

    (iii)   Variac for power control

    (iv)   Condenser for heat removal

    (v)   Provision for measuring flow and temperature rise of
           condenser coolant

    (vi)   Provision for tilting heat pipe

    (vii)   Thermocouples for temperature measurement and associated
           read-out system

    (viii) Thermal insulation.

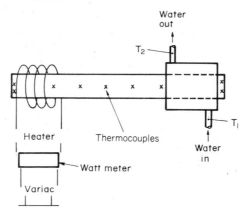

Fig. 4.16   Heat pipe performance test rig

The heater may take several forms, as long as heat is applied uniformly and
the thermal resistance between the heater and evaporator section is low.
This can be achieved using rod heaters mounted in a split copper block
clamped around the heat pipe, or by wrapping insulated heater wire directly
on to the heat pipe.   For many purposes eddy current heating is convenient,
using the condenser as a calorimeter.   Heat losses by radiation and convection to the surroundings should be minimised by applying thermal insulation

to the outside of the heater. An accurate wattmeter covering the anticipated power range, and a variac for close control of power, should be incorporated in the heater circuit. Where orientation may be varied, long leads between the heater and instruments should be used for convenience.

An effective technique for measuring the power output of heat pipes operating at vapour temperatures appropriate to most organic fluids and water is to use a condenser jacket through which a liquid is passed. For many cases this can be water. The heat given up to the water can be obtained if the temperature rise between the condenser inlet and outlet is known, together with the flow rate. The temperature of the liquid flowing through the jacket may be varied to vary the heat pipe vapour temperature. Where performance measurements are required at vapour temperatures of about $0^{\circ}C$, a cryostat may be used.

Cryogenic heat pipes should be tested in a vacuum chamber. This prevents convective heat exchange and a cold wall may be used to keep the environment at the required temperature. As a protection against radiation heat input, the heat pipe, fluid lines and cold wall should all be covered with super-insulation. If the heat pipe is mounted such that the mounting points are all at the same temperature (cold wall and heat sink) it can be assumed that all heat put into the evaporator will be transported by the heat pipe as there will be no heat path to the environment. Further data on cryogenic heat pipe testing can be obtained from references 4.12 and 4.13.

An important factor in many heat pipe applications is the effect of orienta-tion on performance. The heat transport capability of a heat pipe operating with the evaporator below the condenser (thermosiphon or reflux mode) can be up to an order of magnitude higher than that of a heat pipe using the wick to return liquid to the evaporator from a condenser at a lower height. In many cases the wick may prove incapable of functioning when the heat pipe is tilted so that the evaporator is only a few centimeters above the condenser. Of course wick selection is based in part on the likely orientation of the heat pipe in the particular application.

Provision should be made on the rig to rotate the heat pipe through 180° while keeping heater and condenser in operation. The angle of the heat pipe should be accurately set and measured. In testing of heat pipes for

satellites, a tilt of only 0.5 cm over a length of 1 m may be required to
check heat pipe operation, and this requires very accurate rig alignment.

The measurement of temperature profiles along the heat pipe is normally
carried out using thermocouples attached to the heat pipe outer wall. If it
is required to investigate transient behaviour, for example during start-up,
burn-out, or on a variable conductance heat pipe, automatic electronic data
collection is required. For steady state operation a switching box connected
to a digital voltmeter or a multi-channel chart recorder should suffice.

4.3.2 Test procedures.   Once the heat pipe is fully instrumented and set up
in the rig, the condenser jacket flow may be started and heat applied to the
evaporator section. Preferably heat input should be applied at first in steps,
building up to design capability and allowing the temperatures along the heat
pipe to achieve a steady state before adding more power. When the steady
state condition is reached, power input, power output (i.e. condenser flow
rate and ΔT) and the temperature profile along the heat pipe should be noted.

If temperature profiles as shown in Fig. 4.17 are achieved, the heat pipe is
operating satisfactorily. However several modes of failure can occur, all
being recognisable by temperature changes at the evaporator or condenser.

The most common failure is burn-out, created by excessive power input at the
evaporator section. It is brought about by the inability of the wick to feed
sufficient liquid to the evaporator, and is characterised by a rapid rise in
evaporator temperature compared to other regions of the heat pipe. Typically
the early states of burn-out are represented by the upper curve in Fig. 4.17.

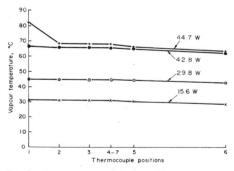

Fig. 4.17 Typical temperature profiles along a heat pipe
under test.

Once burn-out has occurred, the wick has to be reprimed and this is best
achieved by cutting off the power input completely. When the temperature
difference along the pipe drops to 1 - 2°C, the power may be re-applied. The
wick must reprime, i.e. be rewetted and saturated with working fluid along
its complete length, if operation against gravity or in zero gravity is
envisaged. If this is the case, the recovery after burn-out must be demonstrated in the tilted condition. In other cases the recovery may be aided by
gravity assistance.

A second failure mechanism recognisable by an increased evaporator temperature, and known as over-heating, occurs at elevated temperatures. As
explained in Chapters 2 and 3, each working fluid has an operating temperature range characterised by the Merit number, which achieves an optimum
value at a particular temperature and then decreases as this temperature is
exceeded. This means that the fluid is able to transport less heat. Thus
the temperature of the evaporator becomes higher than the rest of the pipe.
In general the evaporator temperature does not increase as quickly as in a
burn-out condition, but these two phenomena are difficult to distinguish.

Temperature changes at the condenser section can also point to failure mechanisms or a decrease in performance. A sudden drop in temperature at the end
of the heat pipe downstream of the cooling jacket occurring at high powers
can be attributed to the collection of working fluid in that region,
insulating the wall and creating a cold spot. This has been called 'cool-
out' (4.14). Complete failure need not necessarily occur when this happens,
but the overall ΔT will be substantially increased and the effective heat pipe
length reduced.

A similar drop in temperature downstream of the condenser jacket can occur in
pipes of small diameter (< 6 mm bore) when the fluid inventory is greater than
that needed to completely saturate the wick. The vapour tends to push the
excess fluid to the cooler end of the heat pipe, where, because of the small
vapour space volume, a small excess of fluid will create a long cold region.
This can occur at low powers and adjustments in fluid inventory may be made
if a valve is incorporated in the heat pipe. One way around this is to use
an excess fluid reservoir, which acts as a sponge but has pores sufficiently
large to prevent it from sucking fluid out of the wick. This technique is

used in heat pipes for space use and the reservoir may be located at any
convenient part of the vapour space.

Failure can be brought about by incompatibilities of materials, generally in
the form of the generation of non-condensable gases which collect in the
condenser section. Unlike liquid accumulation, the gas volume is a function
of vapour temperature and its presence is easily identified.

Unsatisfactory wick cleaning can inhibit wetting, and if partial wetting
occurs the heat pipe will burn out very quickly after the application of even
small amounts of power.

4.3.3 Evaluation testing of a copper heat pipe and typical performance.

Capabilities.   A copper heat pipe using water as the working fluid was manu -
factured and tested, to determine the temperature profiles and the maximum
capability.

The design parameters of the pipe were as follows:

Length	320 mm
Outside dia.	12.75 mm
Inside dia.	10.75 mm
Material of case	copper
Wick form	4 layers 400 mesh
Wick wire dia.	0.025 mm
Effective pore radius	0.031 mm
Calculated porosity	0.686
Wick material	stainless steel
Locating spring length	320 mm
Pitch	7 mm
Wire dia.	1 mm
Material	stainless steel
Working fluid	water (10^6 Ω resistivity)
Quantity	2 ml
End fittings	copper
Instrumentation thermocouples	(7)

Test procedure: The evaporator section was fitted into the 100 mm long heater block in the test rig, and the condenser section covered by a 150 mm long water jacket. The whole system was then lagged.

First tests were carried out with the heat pipe operating vertically with gravity assistance. The power was applied and on achievement of a steady state condition the thermocouple readings and temperature rise through the water jacket were noted, as was the flow rate.

Power to the heaters was increased incrementally and steady state readings noted until dry-out was seen to occur. (This was characterised by a sudden increase in the potential of the thermocouple at the evaporator section relative to the readings of the other thermocouples).

The above procedure was performed for various vapour temperatures and heat pipe orientations with respect to gravity.

Test results: Typical results obtained are shown in Fig. 4.17, showing the vapour temperature profile along the pipe when operating with the evaporator 10 mm above the condenser.

The table below gives power capabilities for a 9.5 mm o.d. copper heat pipe of length 30 cm, with a composite wick of 100 and 400 mesh, operating at an elevation (evaporator above condenser) of 18 cm.

Vapour temp. (^{o}C)	Power out. (W)
84	17
121	30.5
162.5	54
197	89

The working fluid was again water, and a capability of 165 W was measured with horizontal operation. (290 W with gravity assistance).

REFERENCES

4.1   Brown-Boveri & Cie Ag, UK Patent 1281272, April, 1969. (See
       Appendix 7).

4.2   Evans, U.R. The corrosion and oxidation of metals. First Supplem-
       entary Volume, St. Martin's Press, Inc., 1968.

4.3   Organic Solvents; ed. Weissberger, Proskauer, Riddick & Toops;
       Interscience, 1955.

4.4   Birnbreier and Gammel, G. Long time tests of Nb 1% Zr heat pipes
       filled with sodium and caesium. International Heat Pipe Conference
       Stuttgart, October, 1973.

4.5   Grover, G.M., Kemme, J. E and Keddy, E.S. Advances in heat pipe
       technology. International Symposium on Thermionic Electrical Power
       Generation. Stresa, Italy. May, 1968.

4.6   Rice, G.R. and Jennings, J.D. Heat pipe filling. International
       Heat Pipe Conference, Stuttgart, October, 1973.

4.7   Vinz, P., Cappelletti, C. and Geiger, F. Development of capillary
       structures for high performance sodium heat pipes. ibid.

4.8   Quataert, D., Busse, C.A. and Geiger, F. Long time behaviour of
       high temperature tungsten-rhenium heat pipes with lithium and
       silver as the working fluid. ibid.

4.9   Busse, C.A., Geiger, F. and Strub, H. High temperature lithium heat
       pipes. International Symposium on Thermionic Electrical Power
       Generation. Stresa, Italy. May, 1968.

4.10   Baker, E. Prediction of long term heat pipe performance from
       accelerated life tests. AIAA Journal, Vo. 11, No. 9, Sept. 1973.

4.11   Anderson, W.T. Hydrogen evolution in nickel-water heat pipes.
       AIAA Paper 73-726, 1973.

4.12   Kissner, G.L. Development of a cryogenic heat pipe. 1st Int. Heat
       Pipe Conference. Paper 10-2, Stuttgart, Oct. 1973.

4.13   Nelson, B.E. and Petrie, W. Experimental evaluation of a cryogenic
       heat pipe/radiator in a vacuum chamber. 1st Int. Heat Pipe Confer-
       ence. Paper 10-2a, Stuttgart, Oct. 1973.

4.14   Marshburn, J.P. Heat pipe investigations. NASA TN-D-7219. Aug.
       1973.

Special Types of Heat Pipe

References have been made in the text to several types of heat pipe other
than the simple tubular form. It has only been possible to give limited
descriptions of these types of heat pipe so far, and it is considered
valuable to elaborate on their characteristics, and to introduce other
developments, such as electro-osmotic flow pumping, which improve the
performance and widen the application of heat pipes. Variable conductance
heat pipes are discussed in Chapter 6.

The following types will be described:

    (i)   Flat plate heat pipes

    (ii)   Flexible heat pipes

    (iii)   Osmotic heat pipes

    (iv)   Electro-osmotic flow pumping and electro-hydrodynamics

    (v)   Anti-gravity thermosyphon

    (vi)   Thermal diodes and switches

    (vii) Rotating heat pipes

5.1 Flat Plate Heat Pipes

The flat plate heat pipe, one form of which is illustrated in Fig. 5.1,
functions in the same manner as the conventional tubular heat pipe, the main
difference being the form the wick takes to enable liquid distribution over
a wide surface area to be obtained.

The main characteristic of this form of heat pipe is its ability to produce
a surface with very small temperature gradients across it. This near-
isothermal surface can be used to even out and remove hot spots produced by
heaters, or to produce a very efficient radiator for cooling of devices
mounted on it. Also, by mounting a number of heat-generating components on
a flat plate heat pipe, they can be operated at similar temperatures due to
the in-built equalisation process resulting from the fact that the vapour

space will be at a fixed uniform temperature.

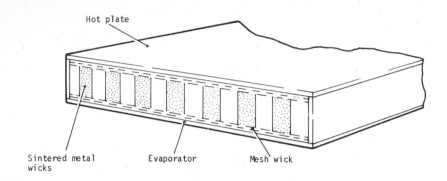

Fig. 5.1   Internal structure of flat plate heat pipe

The wick structure shown in Fig. 5.1 is designed to ensure that liquid can be
returned to and distributed over the top surface of the plate in cases where
operation against gravity is required. If the heat input is via the base,
the sintered structure could be omitted, provision for liquid distribution
across the bottom plate only being necessary.

Location of the evaporator section along one edge of the plate is possible,
and where a limited space is available for the evaporator, the vapour could
be fed into the flat plate vapour chamber and the heat distributed over the
larger flat surface.

The Marconi Company have developed a plate heat pipe which has a conformable
wall constructed using a thin polyester film (see Chapter 7). The flexible
wall of the heat pipe is pressed against the components and the vapour
pressure in the heat pipe during operation ensures that the conformable face
takes up the shape of the components, giving good thermal contact. One
important advantage of this arrangement is that it obviates the need to
incorporate penetrations in the heat pipe to accommodate bolt-on components
or as a passage for electrical leads.

5.2 Flexible Heat Pipes

No work of major significance has been reported on flexible heat pipes during
the past three years. However several manufactureres in the United States

offer these units, and their desirability where source vibration is
encountered, or where difficult assembly problems inhibit incorporation of a
rigid heat pipe, is not in question.

The flexibility may be introduced by incorporating a bellows-type structure
between the evaporator and condenser, or a simple flexible tube of some
plastic material, with conventional metal sections for heat input and
extraction regions.

Bliss et al (5.1) and RCA Corporation (5.2), as well as Eastman (5.3) have
all reported work on flexible heat pipes with hollow bellows structures as
the flexible component.

The system constructed by Bliss, illustrated in Fig. 5.2 was tested while
stationary with varying degrees of bend up to 90°C and while undergoing
transverse and longitudinal vibration in an unbent mode. The heat pipe
performance did undergo a change due to bending and vibration. This change
resulted in an increased maximum horizontal heat transfer capability, and a
reduction in the wick pumping capacity when operating against gravity. It
was also found that critical longitudinal vibrations existed which could
cause the heat pipe to cease operating completely. This occurred at a 10 cps
frequency, when the power input was approaching 500 W.

In order to introduce a degree of flexibility into the wick structure, RCA
proposed and patented a mesh wick woven such that all the wires ran in
transverse directions relative to the longitudinal axis of the heat pipe.

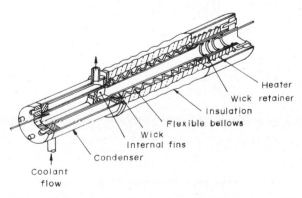

Fig. 5.2 Flexible heat pipe construction (Courtesy RCA Corp)

This, it was claimed, made the pipe more flexible and assisted the prevention of wick mechanical failure. RCA successfully operated a low power heat pipe using a mesh as described above.

Commercially available flexible heat pipes commonly incorporate a plastic section between evaporator and condenser.

5.3 Simple Osmosis

The use of osmosis without the application of any electrical or magnetic field has been proposed by several workers, and is the subject of at least two patents (US Patents 3561525 and 3677337 attributed to S.C. Baer and L.L. Midolo respectively). (See Appendix 7).

Both cite the limitations of capillary action and claim that osmosis could increase substantially the performance of the heat pipe in zero gravity and, perhaps more significantly, against gravity. The embodiment described by Baer, which is essentially similar to that of the later patent given to Midolo, is illustrated in one form in Fig. 5.3.

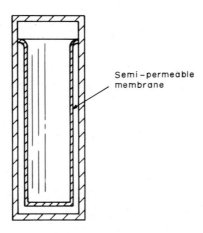

Semi-permeable membrane

Fig. 5.3 Osmotic flow heat pipe invented by Baer
(Courtesy Energy Conversion Systems Inc.)

The containment vessel, as with any heat pipe, is sealed and contains only the working fluid and vapour. Evaporator and condenser sections may be similarly positioned as in a capillary driven system. A return channel for

the liquid is separated from the central vapour flow passage by a semi-permeable membrane such as cellulose. Although not shown, Baer also proposes that
a conventional wick may be used to complement the capillary action. One
working solution suggested is sugar and water, and in order to prevent cross-
contamination of the solvent in the vapour passage and the liquid behind the
membrane, a seal is provided. Application of heat to the working solution
causes it to evaporate to a pure solvent which enters the vapour passage and
flows to the cooler portion of the container, where condensation occurs, onto
the semi-permeable membrane. The pure liquid solvent then passes through the
membrane into solution in the return channel.

The passage of the solvent through the membrane creates an osmotic pressure
considerably stronger than the capillary action created by surface tension.
The pressure is considerably greater than the hydrostatic head of the solution
in the return channel, and a flow of solution to the evaporator is effected.

Midolo claims heat transport capabilities an order of magnitude greater than
that obtained using capillary action, and cites water soluble chlorides,
chlorates and borates as suitable chemicals for forming the working solution.

5.4 Electro-Osmotic Flow Pumping and Electro-hydrodynamics

One method for enhancing liquid pumping capability which has been successfully used in heat pipes is the application of electrokinetics, which
encompasses electro-osmotic flow pumping (5.4, 5.5).

Electrokinetics is the name given to the electrical phenomena which
accompanies the relative movement of a liquid and a solid. These phenomena
are ascribed to the presence of a potential difference at the interface
between any two phases at which movements occur. Thus if the potential is
supposed to result from the existance of electrically charged layers of
opposite sign at the interface, then the application of an electrical field
must result in the displacement of one layer with respect to the other. If
the solid phase is fixed, as in the case of a heat pipe wick, while the
liquid is free to move, the liquid will tend to flow through the pores of
the wick as a consequence of the applied field. This movement is known as
electro-osmosis.

An electro-osmotic heat pipe was designed and tested by Abu-Romia at Brooklyn Polytechnic Institute, New York (5.6). It has the same dimensions as type 'A' used by Cosgrove et al (5.7) in their study, i.e. an overall length of 41 cm and diameter of approximately 5 cm. The wick was formed using glass beads and the working fluids were distilled water in equilibrium with atmospheric carbon dioxide, 10^{-5} HCl, 10^{-5} KCl and 10^{-5} KOH. (10^{-5} denotes concentration in moles per litre).

The electro-osmotic effect was achieved by inserting two porous electrodes, as in Fig. 5.5, and applying a potential across them. Abu-Romia obtained results with a potential of 20 volts, but predicted higher capabilities with higher values of E, although electric heating would begin to affect results at very high potentials.

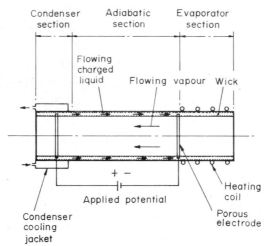

Fig. 5.4 Electro-osmotic wick system
(Courtesy American Institute of Aeronautics and Astronautics and M.M. Abu-Romia)

The highest heat transport capability was achieved with the potassium hydroxide solution, more than doubling the capability, compared with Cosgrove's model, when operating horizontally, and trebling the capability with the evaporator vertically above the condenser. Approximately 2.4 kW was transported in the latter configuration.

It was concluded that the utilisation of electro-osmotic flow pumping in heat pipes could significantly improve performance, particularly when the wicking limit governs the maximum capability, as in operation against gravity. It could also be used to aid start-up and reduce transients. The major limitation was the necessity to adopt wicks and working fluids with a high electrical resistance.

A yet more radical design of heat pipe is that proposed by T.B. Jones (5.8) in which the complete capillary assembly may be replaced by an electrode structure which generates an electrohydrodynamic force. The heat pipe is restricted to the use of insulating dielectric liquids as the working fluid, but as these tend to have a poor wicking capability and can be used in the vapour temperature range between $150^{\circ}C$ and $350^{\circ}C$, where suitable working fluids are difficult to find, performance enhancement is useful.

The electrohydrodynamic heat pipe proposed by Jones was to consist of a thin-walled tube of aluminium or some other good electrical conductor, with end caps made of an insulating material such as plexiglass. A thin ribbon electrode is stretched and fixed to the end caps in such a way that a small annulus is formed between it and the heat pipe wall over the complete length of the heat pipe. (This annulus is only confined to about 20 per cent of the heat pipe circumference, and provision must be made for distributing the liquid around the evaporator by conventional means).

When a sufficiently high voltage is applied, the working fluid collects in the high electric field region between the electrode and the heat pipe wall, forming a type of artery as shown in Fig. 5.5. Evaporation of the liquid causes a nett recession at the evaporator, whereas cooling at the condenser causes an outward bulging of the liquid interface. This creates an inequality in the electromechanical surface forces acting normal to the liquid surface, causing a negative pressure gradiant between condenser and evaporator. Thus a liquid flow is established between the two ends of the heat pipe.

Jones calculated that Dowtherm A could be pumped over a distance approaching 50 cm against gravity, much greater than achievable with conventional wicks. Applications of this technique could include temperature control and arterial priming.

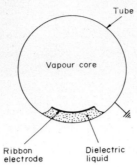

Fig. 5.5 Electrohydrodynamic liquid pump in a heat pipe
(Courtesy International Journal of Heat and Mass Transfer)

The use of an insulating dielectric liquid somewhat limits the applicability of such heat pipes, but within this constraint, performances have shown 'significant improvements' over comparable capillary-wicked heat pipes. Loehrke and Debs (5.22) tested a heat pipe with Freon 11 as the working fluid, and Loehrke and Sebits (5.23) extended this work to flat plate heat pipes, both systems using open grooves for liquid transport between evaporator and condenser. It is of particular interest to note that evaporator liquid supply was maintained even when nucleate boiling was taking place in the evaporator section, corresponding to the highest heat fluxes recorded.

Jones (5.24), in a study of electrohydrodynamic effects on film boiling, found that the pool boiling curve revealed the influence of electrostatic fields of varying intensity, and both the peak nucleate flux and minimum film flux increased as the applied voltage was raised, as shown in Fig. 5.6. Jones concluded that this was a surface hydrodynamic mechanism acting at the liquid-vapour interface.

Because of the limitation in the range of working fluids available, it is unlikely that EHD heat pipes will compete with conventional or other types where water can be used as the working fluid. However Jones and Perry (5.25) suggest that at vapour temperatures in excess of $170^{\circ}C$, where other dielectric fluids are available, advantages may result from the lower vapour pressures accrueing to their use. To date no experiments appear to have been carried out on EHD heat pipes over such a temperature range (e.g. $170 - 300^{\circ}C$), however.

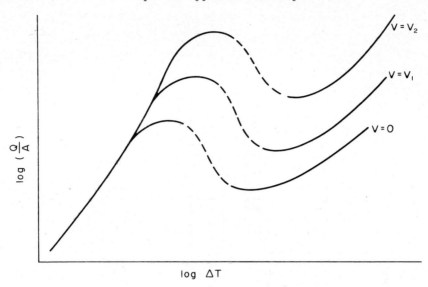

$V = V_2$

$V = V_1$

$V = 0$

log $\left(\dfrac{Q}{A}\right)$

log ΔT

Fig. 5.6. Electrohydrodynamics - the effect of an
electrostatic field on the pool boiling
curve (5.24)

5.5 Anti-Gravity Thermosyphons or Inverse Thermosyphons

In the conventional thermosyphon, described in Chapter 1, the evaporator must
be located below the condenser for satisfactory operation, as the device has
to rely on gravity for condensate return. It is therefore ineffective in zero
gravity or in cases where liquid has to be returned against a gravity head,
however small.

Work carried out at the National Engineering Laboratory (5.9) has led to the
development of what has been called the 'anti-gravity thermosyphon', and the
arrangement is illustrated in Fig. 5.7. As with conventional heat pipes and
thermosyphons, the container is sealed and contains only the working fluid in
liquid and vapour form. In order to drive the condensate back to the evaporator
section, a vapour lift pump is used. This takes the form of a tube with its
base in the sump containing the condensate, the other end having an aperture
leading to the evaporator. The vapour lift pump requires a small heat input at
the base to operate, creating a two-phase mixture in the tube which, having a
lower density than the liquid in the rest of the sump, is forced back up to
the top of the tube (known as the riser tube).

The evaporator takes the form of an annulus sealed at the base, into which

liquid is fed via the vapour pump. An additional annular baffle is provided to
aid circulation in the evaporator. One significant advantage of this device is
that one is not restricted by boiling limitations in the evaporator unlike the
conventional heat pipe.

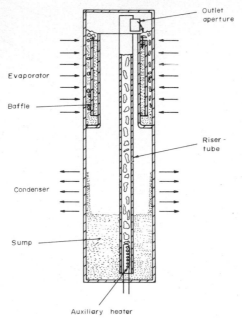

Fig. 5.7 Anti-gravity thermosyphon
(Courtesy National Engineering Laboratory)

Chisholm's paper presents the theory of the anti-gravity thermosyphon, giving
expressions for axial flux in terms of the riser tube diameter and lift capability.

With water as the working fluid, and operating at $100^{\circ}C$ vapour temperature with
riser tube diameter of 1 cm a thermosyphon of 2 cm diameter could support an
axial flux of 1.2 kW/cm^2 with a lift ratio of 10. This increases to 4 kW/cm^2
if the lift ratio is reduced to 5.

Multi-stage vapour lift pumps are also possible, and a three stage unit could
theoretically transport 14 kW over a vertical distance of 10 m, with water at
a vapour temperature of $100^{\circ}C$.

Further concepts of the inverse thermosyphon proposed by the National Engineer-

ing Laboratory are shown in Figs. 5.8 and 5.9. The arrangement in Fig. 5.8 is
a 'symmetrical thermosyphon' in which an evaporator is located above and below
the condenser. No additional heater is required, part of the heating surface
of the lower evaporator being used to drive the vapour lift pump (5.26). The
use of arteries in inverse thermosyphons can be beneficial. In Fig. 5.9(a) the
heated artery allows the liquid sump, shown in Fig. 5.7, to be eliminated. The
driving force for circulation is provided by the downcomer, which runs full
of liquid. The unheated artery concept, shown in Fig. 5.9(b), differs from the
heated artery system in that the vapour instead of being generated in the artery as it passes through the evaporator section, enters through a hole in the
artery between the evaporator and condenser.

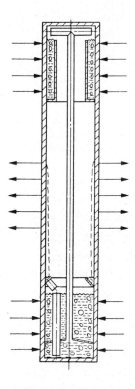

Fig. 5.8   A symmetrical thermosyphon concept (5.26)

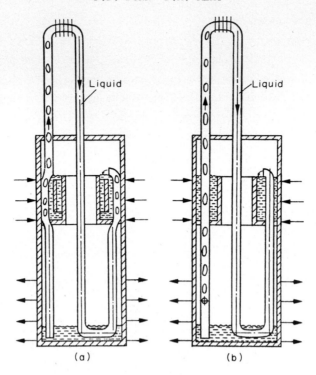

Fig. 5.9   Inverse thermosyphons proposed by NEL with (a)
            heated artery and (b) unheated artery (5.26)

5.6 Heat Pipe Switches and Diodes

In addition to the heat transfer capabilities of the heat pipes which have so
far been discussed, it is possible for heat pipes to be made which act as thermal impedances. These functions greatly increase the range of application of
the heat pipe. For example close temperature control is often essential in
space vehicles whilst the internal dissipation and external heat flux may vary
widely. In cryogenic and other applications heat must be transferred efficient-
ly in one direction but flow in the reverse direction must be avoided.

In order to achieve the control of the heat pipe one or more of the four stages
of the heat pipe must be modified. These stages are:

(i)   Evaporation
(ii)   Vapour flow to condenser
(iii)   Condensation

(iv)   Liquid flow to evaporator

For unidirectional operation (the diode) it is necessary to introduce some
asymmetry into the flow processes. For switching a complete cut off must be
introduced into one of the processes, for example by freezing the working fluid,
removing the working fluid, drying out the wick, interrupting the vapour flow,
or shielding the condenser with inert gas. Variable impedance, or conductance,
is usually achieved by blanking off part of the condenser area using an inert
gas. The effect may be controlled by manually altering the gas pressure, by use
of a large volume inert gas buffer volume which has an effect analogous to
that of a temperature sensitive electrical resistor, or by use of a temperature
sensor and active feedback loop varying the inert gas pressure. The variable
conductance pipe is discussed fully in Chapter 6. Thermal diodes and switches
are considered below.

5.6.1. The thermal diode.

The simplest thermal diode is the thermosyphon in
which gravity provides the asymmetry, but of course with the restriction on
positioning. Gravity will also give a diode effect in the wicked heat pipe
since

$$\Delta P_c = \Delta P_\ell + \Delta P_v \pm \Delta P_g$$

Reversal of direction of flow will reserve the sign of ΔP_g and provided that
$|\Delta P_g| > \Delta P_c$ the pipe will behave as a diode.

Kirkpatrick (5.10) describes two types of thermal diode, one employing liquid
trapping and the second liquid blockage. Referring to Fig. 5.10, with the heat
flow shown in Fig. 5.10(a), the heat pipe will behave normally. If the relative
positions of the evaporator and condenser are reversed then the condensing liquid is trapped in the reservoir whose wick is not connected to the pipe wick on
the left hand side of the diagram and cannot return. The pipe will then not
operate and no heat transfer will occur.

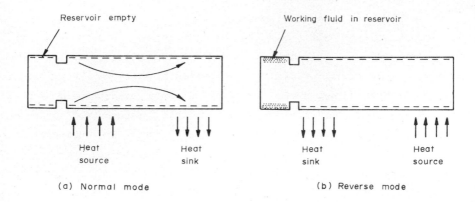

Fig. 5.10 Liquid trap diode

Fig. 5.11 shows a similar arrangement but in this case excess liquid is placed in the pipe. In Fig. 5.11 (a) this liquid will accumulate in the reservoir at the condensing end and the pipe will operate normally. In Fig. 5.11 (b) the positions of the evaporator and condenser are reversed and the excess liquid blocks off the evaporator and the pipe ceases to operate.

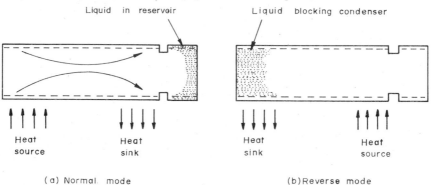

Fig. 5.11 Liquid blockage diode

5.6.2 The thermal switch. A number of methods for switching off the heat pipe have been referred to. Some examples are discussed by Brost and Schubert (5.11) and Eddleston and Hecks (5.12). Fig. 5.12 (a) shows a simple displacement method in which the liquid working fluid can be displaced from an unwicked reservoir by a solid displacer body (5.11). Fig. 5.12 (b) shows interruption of the vapour flow by means of a magnetically operated vane. The working fluid may be frozen by means of a thermoelectric cooler.

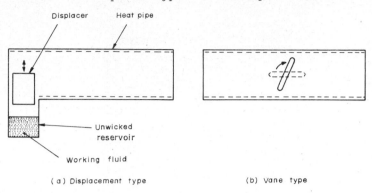

(a) Displacement type   (b) Vane type

Fig. 5.12 Thermal switches

Fig. 5.13 shows such an arrangement due to Rice (5.13)

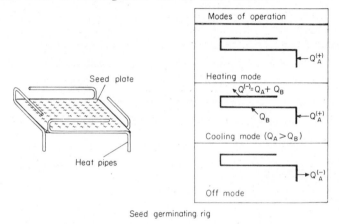

Seed germinating rig

Fig 5.13 Thermal switch application developed
at Reading University

5.7 The Rotating Heat Pipe

The rotating heat pipe is a two phase thermosyphon in which the condensate is
returned to the evaporator by means of centrifugal force. The rotating heat
pipe consists of a sealed hollow shaft, having a slight internal tape along
its axial length, and containing a fixed amount of working fluid, Fig. 5.14.

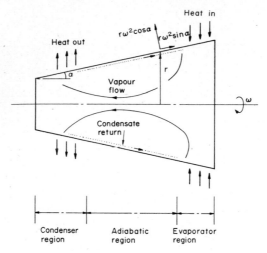

Fig. 5.14   Rotating heat pipe

The rotating heat pipe, like the conventional capillary heat pipe, is divided
into three sections, the evaporator region, the adiabatic region and the condenser region. However, the rotation about the axis will cause a centrifugal
acceleration $\omega^2 r$ with a component $\omega^2 r \sin \alpha$ along the wall of the pipe. The
corresponding force will cause the condensed working fluid to flow along the
wall back to the evaporator region.

The first reference to the rotating heat pipe was given in an article by Gray
(5.14).

Centrifugal forces will significantly affect the heat and mass transfer effects
in the rotating heat pipe and the three regions will be considered in turn.

Published work on evaporation from rotating boilers, Gray et al (5.15), suggests
that high centrifugal accelerations have smooth, stable interfaces between the
liquid and vapour phases. Using water at a pressure of one atmosphere and centrifugal accelerations up to 400 g heat fluxes of up to 257 W/cm^2 were obtained
The boiling heat transfer coefficient was similar to that at 1 g, however the
peak, or critical flux, increased with acceleration. Costello and Adams (5.16)
have derived a theoretical relationship which predicts that the peak heat flux
increases as the one fourth power of acceleration.

In the rotating condenser region a high condensing coefficient is maintained

due to the efficient removal of the condensate from the cooled liquid surface
by centrifugal action. Ballbach (5.17) has carried out a Nusselt-type analysis
but neglected vapour drag effects. Daniels and Jumaily (5.18) have carried out
a similar analysis but have taken account of the drag force between the axial
vapour flow and the rotating liquid surface. They concluded that the vapour
drag effect was small and could be neglected except at high heat fluxes. These
workers also compared their theoretical predictions with measurements made on
rotating heat pipes using Arcton 113, Arcton 21 and water as the working fluid.
They stated that there is an optimum working fluid loading for a given heat
pipe geometry, speed and heat flow. The experimental results appear to verify
the theory over a range of heat flow and discrepancies can be explained by experimental factors. An interesting result (5.19) from this work is the establishment of a figure of merit M' for the working fluid where:

$$M' = \frac{\rho_\ell^2 L k_\ell^3}{\mu_\ell}$$

ρ_ℓ is the liquid density

L   is the latent heat of vaporisation

k_ℓ is the liquid thermal conductivity

μ_ℓ is the liquid viscosity

M' is plotted against temperature for a number of working fluids in Fig. 5.15.

Normally the rotating heat pipe should have a thermal conductance comparable
to or higher than that of the capillary heat pipe. The low equivalent conductance quoted by Daniels and Jumaily (5.18) may have been due to a combination
of very low thermal conductivity of the liquid Arcton and a relatively thick
layer in the condenser.

In the adiabatic region, as in the same region of the capillary heat pipe, the
vapour and liquid flows will be in the opposite directions, with the vapour
velocity much higher than the liquid.

5.7.1 Factors limiting the heat transfer capacity of the rotating heat pipe.

The factors which will set a limit to the heat capacity of the rotating heat

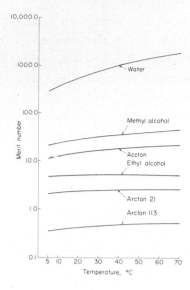

Fig. 5.15 Property group (figure of merit) versus temperature for various working fluids in a rotating heat
pipe.

pipe will be sonic, entrainment, boiling and condensing limits (and non-condensable gases). The sonic limit and non-condensable gas effects are the same as
for the capillary heat pipe. Entrainment will occur if the shear forces due to
the counter flow vapour are sufficient to remove droplets and carry them to the
condenser region. The radial centrifugal forces are important in inhibiting
the formation of the ripples on the liquid condensate surface which precede
droplet formation.

The affect of rotating on the boiling limit has already been referred to (5.15,
5.16) as has the condensing limit (5.18).

5.7.2 Applications of rotating heat pipes. The rotating heat pipe is obviously applicable to rotating shafts having energy dissipating loads, for example
the rotors of electrical machinery, rotary cutting tools, heavily loaded bear-
ings and the rollers of presses. Polasek (5.20) reports experiments on cooling
an a.c. motor incorporating a rotating heat pipe in the hollow shaft, Fig. 5.16.
He reported that the power output can be increased by 15% with the heat pipe
without any rise in winding temperature. Gray (5.14) suggested the use of a
rotating heat pipe in an air conditioning system. Groll et al (5.21) reported
the use of a rotating heat pipe for temperature flattening of a rotating drum.

The drum was used for stretching plastic fibres and rotated at 4000 - 6000 rev/min, being maintained at a temperature of 250°C. Groll selected diphenyl as the working fluid.

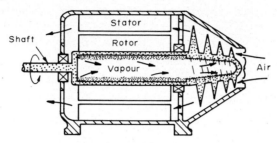

Fig. 5.16 Application of rotating heat pipes to cooling
           of motor rotors.

5.7.3 **Most recent work.** The 2nd International Heat Pipe Conference provided a forum for reports of several more theoretical and experimental studies on rotating heat pipes.

Marto (5.27) reported on recent work at the Naval Post-graduate School, Monterey, where studies on this type of heat pipe have been continuing since 1969. The latest developments at Monterey concern the derivation of a theoretical model for laminar film condensation in the rotating heat pipe, taking into account vapour shear and vapour pressure drop. Marto found that these effects were small, but he recommended that the internal condensation resistance be of the same order as the condenser wall resistance and outside convection resistance. Thus it is desirable in rotating heat pipes to make the wall as thin as possible, although in rotating electrical machines whis is often inconsistent with structural requirements.

Vasiliev and Khrolenok (5.28) extended the analysis to the evaporator section, and recommended that:

(i)   The most favourable mode of evaporator operation is fully developed
        nucleate boiling with minimal free convection effects because high
        heat transfer coefficients are obtained, independent of liquid film
        thickness and rotational speed.

(ii)   For effective operation when boiling is not fully developed, the
        liquid film, preferably as thin as possible, should cover the whole
        evaporator surface.

HP—H

(iii)   Condenser heat flux is improved with increasing rotational speed.
         (For a 400 mm long x 70 mm internal diameter unit, with water as the
         working fluid, Vasiliev achieved condenser coefficients of 5000
         $W/m^2\,^{\circ}C$ at 1500 rev/min with an axial heat transport of 1600 W).

Daniels and Williams (5.29), in an extension of the work described above, detailed theoretical and experimental work on the effect of non-condensable gas
on rotating heat pipe performance. A marked reduction in heat transport capability was noted, resulting directly from a lowering of the saturation temperature at the condensate/gas interface.

REFERENCES

5.1   Bliss, F.E. et al. Construction and test of a flexible heat pipe. ASME
       Paper 70-HT/SpT-13, 1970.

5.2   RCA Corporation. Flexible heat pipe. UK Patent 1322276, published 4 July
       1973.

5.3   Eastman, G.Y. The heat pipe - a progress report. 4th Intersoc. Energy
       Conversion Conf., Washington D.C., Sept. 1969, pp. 873 - 878 AIChE,
       1969.

5.4   Dresner, L. Electrokinetic phenomena in charged microcapillaries, J.
       Phys. Chemistry, Vol. 67, p. 1635, 1963.

5.5   Burgeen, D. and Nakache, F.R. Electrokinetic flow in ultrafine capillary
       slits. J. Physical Chemistry, Vol. 68, p. 1084, 1964.

5.6   Abu-Romia, M.M. Possible application of electro-osmotic flow pumping in
       heat pipes. AIAA Paper 71-423, 1971.

5.7   Cosgrove, J.H. et al. Operating characteristics of capillary-limited
       heat pipes. J. Nuclear Energy, Vol. 21, pp 547 - 558, Pergamon Press,
       1967.

5.8   Jones, T.B. Electrohydrodynamic heat pipes. Int. J. Heat and Mass Trans-
       fer, Vol. 16, pp 1045 - 1048, 1973.

5.9   Chisholm, D. The anti-gravity thermosyphon. Symposium on Multiphase Flow
       Systems. I. Mech. E./I. Chem. E., Symposium Series 38, 1974.

5.10   Kirkpatrick, J.P. Variable conductance heat pipes - from the laboratory
       to space. 1st International Heat Pipe Conference, Stuttgart, Oct. 15 - 17,
       1973.

5.11   Brost, O. and Schubert, K.P. Development of alkali-metal heat pipes as
       thermal switches. ibid.

5.12   Eddleston, B.N.F. and Hecks, K. Application of heat pipes to the thermal
       control of advanced communications spacecraft. ibid.

5.13   Rice, G. Four-way thermogradient plate for seed germinating studies.
       Reading University, 1975.

5.14   Gray, V.H. The rotating heat pipe. A wickless hollow shaft for transfer-
       ring high heat fluxes. ASME Paper No. 69-HT-19, 1969.

5.15 Gray, V.H., Marto, P.J. and Joslyn, A.W. Boiling heat transfer coeffici-
      ents: interface behaviour and vapour quality in rotating boiler operation
      to 475 g. NASA TN D-4136, March 1968.

5.16 Costello, C.P. and Adams, J.M. Burn out fluxes in pool boiling at high
      accelerations. Mech. Eng. Dept., University of Washington, Washington
      D.C., 1960.

5.17 Ballback, L.J. The operation of a rotating wickless heat pipe. M.Sc.
      Thesis. Monterey, Calif., United States Naval Postgraduate School, 1969
      (AD 701 674).

5.18 Daniels, T.C. and Al-Jumaily, F.K. Theoretical and experimental analysis
      of a rotating wickless heat pipe. Proc. 1st International Heat Pipe Con-
      ference, Stuttgart, Oct. 1973.

5.19 Al-Jumaily, F.K. An investigation of the factors affecting the perform-
      ance of a rotating heat pipe. Ph.D. Thesis, University of Wales, Dec.
      1973.

5.20 Polasek, F. Cooling of a.c. motor by heat pipes. Proc. 1st. Internation-
      al Heat Pipe Conference, Stuttgart, Oct. 1973.

5.21 Groll, M., Kraus, G., Kreel, H. and Zimmerman, P. Industrial applications
      of low temperature heat pipes. ibid.

5.22 Loehrke, R.I. and Debs, R.J. Measurements of the performance of an elec-
      trohydrodynamic heat pipe. AIAA Paper 75-659, 10th Thermophys. Conf.,
      1975.

5.23 Loehrke, R.I. and Sebits, D.R. Flat plate electrohydrodynamic heat pipe
      experiments. Proc. 2nd Int. Heat Pipe Conference, Bologna; ESA Report
      SP 112, 1976.

5.24 Jones, T.B. Electrohydrodynamic effects on minimum film boiling. Report
      PB-252 320, Colorado State University, 1976.

5.25 Jones, T.B. and Perry, M.P. Electrohydrodynamic heat pipe experiments.
      J. Appl. Physics, Vol. 45, No. 5, 1974.

5.26 Grant, I.D.R. Inverse thermosyphons. Proc. Heat Pipe Forum. NEL Report
      No. 607, Dept. Industry, Jan. 1976.

5.27 Marto, P.J. Performance characteristics of rotating, wickless heat pipes.
      Proc. 2nd Int. Heat Pipe Conference, Bologna; ESA Report SP 112, 1976.

5.28 Vasiliev, L.L. and Khrolenok, V.V. Centrifugal coaxial heat pipes. ibid.

5.29 Daniels, T.C. and Williams, R.J. Theoretical and experimental analysis
      of non-condensable gas effects in a rotating heat pipe. ibid.

The Variable Conductance Heat Pipe

Brief reference has already been made to the variable conductance heat pipe
(VCHP). The unique feature which sets the VCHP apart from other types of
heat pipe is its ability to maintain a device mounted at the evaporator at a
near constant temperature, independent of the amount of power being generated
by the device.

The temperature control functions of a gas buffered heat pipe were first
examined as a result of non-condensable gas generation within a sodium/
stainless steel basic heat pipe. It was observed (6.1) that as heat was put
into the evaporator section of the heat pipe, the hydrogen generated was
swept to the condenser section. An equilibrium situation shown in Fig. 6.1
was reached.

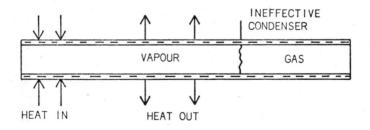

Fig. 6.1 Equilibrium state of a gas-loaded heat pipe

Subsequent visual observation of high temperature heat pipes, and temperature
measurements, indicated that the working fluid vapour and the non-condensable
gas were segregated, that a sharp interface existed between the working fluid
and the non-condensable gas, and that the non-condensable gas effectively
blocked off the condenser section it occupied, stopping any local heat
transfer.

Significantly, it was also observed that the non-condensable gas interface
moved along the pipe as a function of the thermal energy being transported by

the working fluid vapour, and it was concluded that suitable positioning of
the gas interface could be used to control the temperature of the heat input
section within close limits.

Much of the subsequent work on heat pipes containing non-condensable or inert
gases has been in developing means for controlling the position of the gas
front, and in ensuring that the degree of temperature control achievable is
sufficient to enable components adjacent to the evaporator section to be
operated at essentially constant temperatures, independent of their heat
dissipation rates, over a wide range of powers.

The first extension of the simple form of gas buffered pipe shown in Fig. 6.1
was the addition of a reservoir downstream of the condenser section
(Fig. 6.2). This was added to allow all the heat pipe length to be effective
when the pipe was operating at maximum capability, and to provide more
precise control of the vapour temperature. The reservoir could also be
conveniently sealed using a valve.

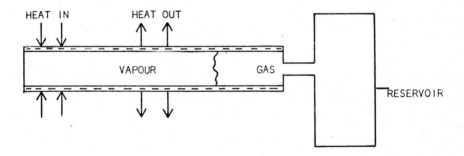

Fig. 6.2   Cold reservoir variable conductance heat pipe

The early workers in the field of cold-reservoir VCHP's were troubled by
vapour diffusion into the reservoir, followed by condensation, even if liquid
flow into the gas area had been arrested. It is necessary to wick the
reservoir of a cold-reservoir unit in order to enable the condensate to be
removed. The partial pressure of the vapour in the reservoir will then be at
the vapour pressure corresponding to its temperature (Fig. 6.3).

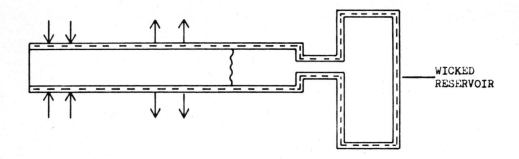

Fig. 6.3 Cold wicked reservoir VCHP

The type of VCHP described above is of the passively controlled type. The
active condenser length varies in accordance with temperature changes in
various parts of the system. An increase in evaporator temperature causes
an increase in vapour pressure of the working fluid, which causes the gas
to compress into a smaller volume, releasing a larger amount of active
condenser length for heat rejection. Conversely, a drop in evaporator
temperature results in a lower vapour pressure, which allows the gas to
expand, shutting off active condenser surface area. The nett effect is to
provide a passively controlled variable condenser area which increases or
decreases heat transfer in response to the heat pipe vapour temperature.

One of the most recent cold reservoir VCHP's was that constructed by Kosson
et al (6.2). An arterial wick system of high liquid transport capability
was used in conjunction with ammonia to carry up to 1200 W. The nominal
length of the heat pipe, including reservoir, approached 2 m and the diameter
was 25 mm. Nitrogen was the control gas.

An important feature of this heat pipe was provision for subcooling the
liquid in the artery. This was to reduce inert gas bubble sizes and helped
the liquid to absorb any gas in the bubbles. This phenomenon is discussed in
detail in Section 6.8

6.1 Passive Control Using Bellows

Wyatt (6.1) proposed as early as 1965 that a bellows might be used to control
the inert gas volume, but he was not specific in suggesting ways in which

the bellows volume might be adjusted.

It is impractical to insert a wick into a bellows unit, and consequently it
is necessary to have a semi-permeable plug between the condenser section and
the bellows in order to prevent the working fluid from accumulating within
the storage volume. The plug must be impervious to both the working fluid
vapour and liquid, but permeable to the non-condensable gas. Marcus and
Fleischman (6.3) proposed and tested a perforated Teflon plug with success,
preventing liquid entering the reservoir during vibration tests.

However, Wyatt did put forward proposals which would have the effect of overcoming one of the major problems associated with the 'cold reservoir' VCHP,
although he did not appreciate the significance at the time. His proposal
was to electrically heat the bellows, which would be thermally insulated from
the environment. His argument for doing this was to ensure that stray
molecules of working fluid would not condense in the bellows if the bellows
temperature was maintained at about 1°C above the heat pipe vapour temperature. However, by controlling the temperature of the non-condensable gas
reservoir, one is able to eliminate the most undesirable feature of the basic
cold reservoir VCHP, namely the susceptibility of the gas to environmental
temperature changes, which can upset the constant temperature performance.

Turner (6.4) investigated the use of bellows to change the reservoir volume
and/or pressure. He proposed a mechanical positioning device to control the
bellows between two precisely determined points, but listed several
disadvantages of this type of control, including the fact that mechanical
devices require electrical energy for their activation, and are also subject
to failure due to jamming and friction. In proposing ammonia as the working
fluid, he also felt that the associated pressure might add considerably to
the weight of the bellows for containment reasons; also fatigue failure was
a possibility.

6.2 Hot Reservoir Variable Conductance Heat Pipes

The cold reservoir VCHP is particularly sensitive to variations in sink
temperature which could affect the reservoir pressure and temperature. In an
attempt to overcome this drawback, the hot-reservoir unit was developed.

One attractive layout for a hot reservoir system is to locate the gas
reservoir adjacent to, or even within, the evaporator section of the heat
pipe. Thermal coupling of the reservoir to the evaporator minimises gas
temperature fluctuations, which limit the controllability. Fig. 6.4
illustrates one form this concept might take.

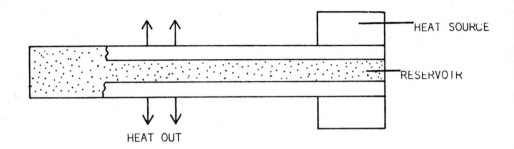

HEAT SOURCE

RESERVOIR

HEAT OUT

Fig. 6.4 Hot reservoir VCHP

It has previously been stated that it is undesirable to have working fluid
inside the hot reservoir, and the semi-permeable plug has been put forward as
one way of preventing diffusion of large amounts of vapour or liquid into the
gas volume. If the reservoir has a wick, and contains working fluid, there
will be within it a vapour pressure corresponding to its temperature, which
in the case of the hot reservoir pipe having a gas reservoir within the
evaporator, would be essentially the same as the temperature throughout the
whole pipe interior, and there would not be gas in the reservoir.

An alternative technique (6.5) applied to the hot-reservoir VCHP is to
provide a cold-trap between the condenser and the reservoir. This effectively reduces the partial pressure of the working fluid vapour in the gas,
and the system provides temperature control which is relatively independent
of ambient radiation environments. This system is not applicable to a hot
reservoir located inside the evaporator.

6.3 Feedback Control Applied to the Variable Conductance Heat Pipe

The cold reservoir VCHP is particularly sensitive to variations in sink temperature which could affect the reservoir pressure and temperature. In an
attempt to overcome this drawback, the hot reservoir unit was developed, as

described above.

Ideally, each of these forms of heat pipe is at best capable of maintaining
its own temperature constant, and this is true only if an infinite storage
volume is used. Thus if the thermal impedance of the heat source is large,
or if the power required to be dissipated by the component is liable to
fluctuations over a range, the temperature of the source would not be kept
constant and severe fluctuations could occur, making the system unacceptable.

The development of feedback controlled variable conductance heat pipes has
enabled absolute temperature control to be obtained, and this has been
demonstrated experimentally (6.6). These heat pipes are representative of
the third generation of thermal control devices incorporating the heat pipe
principal.

Two forms of feedback control are feasible, active (electrical) and passive
(mechanical).

6.3.1 Electrical feedback control (active).

An active feedback controlled
VCHP is shown diagrammatically in Fig. 6.5. A temperature sensor, electronic
controller, and a heated reservoir (internal or external heaters) are used to
adjust the position of the gas-vapour interface such that the source temperature remains constant. As in the cold reservoir system, the wick is
continuous and extends into the storage volume. Consequently, saturated
working fluid is always present in the reservoir. The partial pressure of
the vapour in equilibrium with the liquid in the reservoir is determined by
the reservoir temperature, which can be varied by the auxiliary heater.

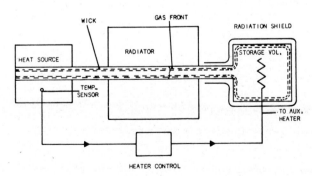

Fig. 6.5 Active feedback controlled VCHP

The two extremes of control required in the system are represented by the
high power/high sink and low power/low sink conditions. The former case
necessitates operation at maximum conductance, conversely the low power/low
sink condition is appropriate to operation with minimum power dissipation.

By using a temperature sensing device at the heat source, and connecting this
via a controller to a heater at the reservoir, the auxiliary power and thus
the reservoir temperature can be regulated so that precise control of the gas-
buffer interface occurs, maintaining the desired source temperature at a
fixed level.

6.3.2 Mechanical feedback control (passive).

Most of the work on mechanical
feedback control has involved the use of a bellows reservoir, as advocated in
several earlier proposals on non-feedback controlled passive variable conductance heat pipes. A proposed passive feedback system utilising bellows was
designed by Bienert et al (6.7, 6.8) and is illustrated in Fig. 6.6. The
control system consists of two bellows and a sensing bulb located adjacent to
the heat source. The inner bellows contains an auxiliary fluid, generally
an incompressible liquid, and is connected to the sensing bulb by a capillary
tube.

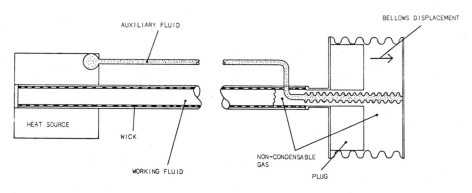

Fig. 6.6 Passive feedback controlled VCHP

Variations in source temperature will cause a change in the pressure of the
auxiliary fluid, resulting in a displacement of the inner bellows. This
displacement causes a movement of the main reservoir bellows. By relating
the displacement of the bellows system, and therefore that of the vapour/gas
interface in the heat pipe, to the heat source, a feedback controlled system

which regulates the source temperature is obtained.

The construction and testing of a passive control VCHP using methanol as the
working fluid and nitrogen as the gas has been reported by Depew et al (6.9).
The system was run over a power input range of 2 - 30 W, with the heat source
at ambient temperature and the heat sink at a nominal value of $0^{\circ}C$. Control
was obtained with a metal bellows gas reservoir which was actuated by an
internal liquid-filled bellows. The liquid bellows was pressurised by
expanding liquid methanol which was contained in an auxiliary reservoir in
the evaporator heater block. Temperature variation of the heat source was
restricted to $\pm 4^{\circ}C$ using this design.

6.4 Other Control Techniques

All of the control methods discussed above use a vapour/gas interface to vary
the area available for heat rejection from the heat pipe.

Other techniques are available and are reviewed in the literature (6.10, 6.11
6.12). Use of valves to throttle vapour flow was proposed in an early study
by Anand (6.13). The amount of control possible using this technique is
limited. Firstly the thermal resistance of the working fluid is only a small
part of the overall resistance in a heat pipe system. Secondly, the pressure
drop in the vapour can be varied over only a limited range without exceeding
the capillary pumping capability of the pipe, except at a low absolute
pressure (Schlosinger 6.14). On-off control can be effected using vapour
blockage, but conduction through the wick and wall can still occur.

Wick discontinuities have been proposed for control but suitable means for
varying the magnitude of the discontinuity would be difficult to implement.
Proposals have also been put forward for blocking the reservoir with excess
liquid. (See also Chapter 5.)

Katzoff (6.15) proposed a high resistance **wick** with an additional artery
system, the latter being thermostatically controlled using a bellows in such
a way that the bellows contracted to plug the artery when the temperature
at the evaporator fell below a certain value.

More speculative techniques put forward include freeze-thaw control, in which

the working fluid is chosen to commence operating at a certain evaporator
temperature level.

6.5 Comparison of Systems

The several systems are compared in Table 6.1. Comparing the two types of
feedback control (active and passive) better temperature control is obtained
using an active system. In the active system all the non-condensable gas
will be in the condenser when the low power sink condition is attained.
However, in the passive arrangement, non-condensable gas will be present in
the reservoir regardless of the use of a plug, and the excess gas must be
accommodated when the bellows is at maximum size (high power/sink condition).
Hence, storage requirements will be generally greater for a passive system
having the same degree of temperature control as an active system. Better
temperature control can be achieved with an active system than with the
equivalent volume passive system. The necessity to incorporate a semi-
permeable plug, and the use of moving parts in the form of a bellows also
adds to the complexity of passive systems.

In some applications the use of additional electrical power to heat the
reservoir, or the increased complication of a bellows system, may be
unacceptable. In this case the choice lies between simple hot or cold
reservoir variable conductance heat pipes.

The selection procedure may be illustrated using the following spacecraft
cooling problem:-

Example: Present initial comparisons between hot and cold reservoir VCHP's
capable of starting up at -100°C and operating with an evaporator temperature
of 35°C and a control requirement of $\pm 5^{\circ}$C. Minimum sink temperature is
-100°C.

6.5.1 Cold reservoir heat pipes. The general advantages and drawbacks are
given in Table 6.1.

Condenser geometry: The condenser volume in a cold reservoir heat pipe
should be as small as possible to ensure that the size of the reservoir is
also small.

TABLE 6.1   COMPARISON OF TEMPERATURE CONTROLLED HEAT
PIPE SYSTEMS

System	Advantages	Disadvantages
Wicked cold reservoir Passive control.	Reliable without moving parts. No auxiliary power needed. Sensitivity is governed by vapour pressure within the reservoir.	Very sensitive to sink conditions. Large storage volume. Only heat pipe temperature controlled
Nonwicked hot reservoir. Passive control.	Reliable without moving parts. Less sensitive to sink conditions than cold reservoir. No auxiliary power needed.	Sensitive to heat carrier diffusion into the reservoir. Only heat pipe temperature controlled.
Passive feedback controlled bellows system.	Heat source control. Slight sensitivity to sink conditions. No auxiliary power needed.	Complex and expensive system. Sensitive to heat carrier diffusion into the reservoir. Moving parts used.
Active electrical feedback controlled system	Heat source control. Best adjustability to various set points. Sensitivity is governed by vapour pressure within the reservoir. Minimum storage volume of all concepts. Relative insensitivity to gas generation. No moving parts.	Auxiliary power needed.

Working fluid: The normal selection criteria for standard heat pipes (i.e. large Merit number, compatibility, etc.) all apply to gas controlled heat pipes. High gas control sensitivity factors should be exhibited by selected working fluids.

The gas control sensitivity factor is a function of vapour temperature and for a fixed temperature may be expressed as $\frac{M L}{R}$ where M is molecular weight of the working fluid, L the latent heat of evaporation and R the universal gas constant. For fluids which are liquid at $-100^{\circ}C$, the values of $\frac{M L}{R}$ are as follows:

(Values are calculated for vapour temperatures of $15^{\circ}C$)

Ethanol	8960 (best)
n-Pentane	5860
Freon 11	5450
Freon 21	5350
Freon 12	4300

Minimum power condition: This condition arises when the heat flow to the condenser is by axial conduction and vapour diffusion, i.e. when the heat pipe is turned off. It is desirable to keep this power leak as low as possible by ensuring that the reservoir has sufficient volume to store gas to completely block off the condenser length. Also a low conductivity region in the wall at this section, i.e. just above the condenser, may be provided. (A stainless steel wall would be more desirable than an aluminium wall on this basis).

Cold traps: The degree of control possible with a cold reservoir VCHP is limited by the large variations in reservoir temperature which may occur. One way in which the reservoir temperature can rise is by axial conduction from the condenser, particularly when maximum heat is being dissipated. This can be minimised by providing a low axial conduction section at the reservoir exit.

Constructional materials: Operation at minimum and maximum power conditions calls for a wall and wick having a low axial thermal conductivity. However, the diffusion freeze-out rate is reduced if such materials are used, and a compromise must be reached.

Transient performance: The transient performance of a cold reservoir heat pipe is a function of the system thermal capacity. The system exhibits good response to power input changes, but unfortunately the response to changes in environmental changes at the reservoir is also rapid.

6.5.2 Hot reservoir heat pipes. The hot reservoir VCHP has a poor response when compared with the cold reservoir system. However, it is able to offer considerably superior control capability because the reservoir, which is not wicked, is located in an essentially constant temperature environment adjacent to or in the heat pipe evaporator section.

Most of the features discussed in relation to the selection of cold reservoir VCHP's can be applied to hot reservoir units. The main difference between the two systems is in transient performance.

Transient performance: The partial pressure of the vapour in a non-wicked reservoir is governed by the diffusion of vapour to and from the end of the condenser, rather than from the reservoir walls. Hence diffusion paths are much longer and the diffusion rate can dictate the transient behaviour of a hot reservoir system.

If, during for example, launch of the vehicle, the liquid in the heat pipe wick is forced out and collects in the reservoir, the heat pipe could take a considerable time to return to normal operating conditions. Until such conditions are reached, the reservoir vapour pressure will be too high, resulting in excessive condenser non-condensable blockage and thus high operating temperatures. Diffusion is the only process by which the excess vapour in the reservoir can be returned to the wick in the condenser section.

If the build-up of vapour pressure in the reservoir is large, as with a high vapour pressure working fluid, the performance of the heat pipe can be radically affected. Also diffusion rates vary inversely with system pressure, and both of these considerations suggest that a low vapour pressure working fluid is best for good response. Also this minimises the need for an extra thick wall to cater for over-pressurisation should excess liquid enter the hot reservoir.

Of the fluids with freezing points below $-100^{\circ}C$, ethanol has the lowest vapour vapour pressure at $35^{\circ}C$ (0.07 bar). The Freons (Arctons) tend to be high, and n-Pentane lies between these extremes (1 bar).

Reservoir sizing - hot and cold reservoirs: While it is not desirable to use the flat front model, where the interface between the vapour and the gas is assumed flat, with no axial conduction or diffusion, this approach can be used to obtain a first order estimate of the various parameters involved.

Marcus (6.16) presents equations for assessing the reservoir volume requirements as a function of the degree of temperature control required. These equations are developed for both hot and cold reservoir heat pipes and are given below. Full condenser utilisation is assumed.

Hot reservoir (non-wicked):

$$\frac{V_c}{V_R} = \left(\frac{P_{va\ max} - P_{vs\ max}}{P_{va\ min} - P_{vs\ min}} \times \frac{T_{s\ min}}{T_{va\ max}} - \frac{T_{s\ min}}{T_{va\ min}} \right)$$

for cases where the reservoir is coupled to the evaporator and $0 \leq \ell_a \leq \ell_c$.

Cold reservoir (wicked):

$$\frac{V_c}{V_R} = \left(\frac{P_{va\ max} - P_{vs\ max}}{P_{va\ min} - P_{vs\ min}} \times \frac{T_{s\ min}}{T_{s\ max}} - 1 \right)$$

for case where the reservoir temperature equals the sink temperature.

Where V_c = condenser volume

V_R = reservoir volume

$P_{va\ max}$ = max. vapour pressure in active zone

$P_{va\ min}$ = min. vapour pressure in active zone

P_{vs} = vapour pressure at sink

$T_{s\ max}$ = max. sink temperature

$T_{s\ min}$ = min. sink temperature

$T_{va\ max}$ = max. temperature in active zone

$T_{va\ min}$ = min. temperature in active zone

ℓ_a = active length

ℓ_c = condenser length (= ℓ_a in high power case)

While these equations are for specific cases, hot non-wicked reservoirs will often be coupled to the evaporator as a stable temperature zone and cold wicked reservoirs will frequently see the same environment as the condenser. These equations can therefore be used as representative of typical VCHP configurations on space vehicles.

Preliminary calculations may be carried out assuming an evaporator temperature of $35^\circ C$ and a control requirement of $\pm 5^\circ C$, with a minimum sink
temperature of $-100^\circ C$.

Results for three working fluids used in both hot non-wicked reservoir and
cold wicked reservoir heat pipes are summarized below.

Working Fluid	V_R/V_c Cold Reservoir	V_R/V_c Hot Reservoir
Ethanol	6.25	2.74
Freon 11	-	5.13
n-Pentane	-	4.35

The only fluid of the above three which could be used in a cold reservoir
VCHP to meet the required evaporator temperature control is ethanol.

All of the fluids listed above could meet our original specification in a hot
reservoir VCHP, but the most compact reservoir would be obtained using
ethanol. Reservoirs for the other fluids would be almost twice as large,
causing weight penalties.

It can be seen that the hot reservoir system is superior in the above example,
and of the fluids proposed, ethanol is the best.

The analysis of the more sophisticated active feedback system is given below.

6.6 Analysis of Feedback Controlled Variable Conductance Heat Pipes

The source and vapour temperature of any heat pipe are related by:

$$T_s = T_v + R_s Q \qquad \qquad \dots 6.1$$

where T_s is the source temperature

T_v is the vapour temperature

Q is the heat load

and R_s is the total heat transfer resistance between the
heat source and the vapour space.

From eqn. 6.1 it is evident that the source temperature will vary with Q as

$$\frac{d\,T_s}{d\,Q} = \frac{d\,T_v}{d\,Q} + R_s \qquad\qquad \dots 6.2$$

In the self controlled heat pipe $\frac{d\,T_v}{d\,Q}$ may approach zero (very large storage volume) but may never be negative. Hence, if the thermal resistance between the source and the pipe is small then the source temperature will remain nearly constant. However, in most cases this is not the case and the source temperature will vary with variation in heat load even though the pipe maintains a constant vapour temperature.

However, when external control is applied $\frac{d\,T_v}{d\,Q}$ can be made negative and the source temperature held constant with varying heat load.

Other problems, as discussed above, which occur with conventional systems include sensitivity to variation in storage temperature (cold reservoirs) and diffusion of the vapour into the non-condensable gas with corresponding excursion of the source temperature. These are also reduced significantly by the use of feedback control.

6.6.1 **Steady state analysis.** The study carried out by Bienert (6.10) has been modified (6.6, 6.7) to include the effect of active and passive feedback control. The main conclusions of this study were:

(i)   Both active and passive feedback control variable conductance heat pipes are feasible and can be made stable.

(ii)   A marked improvement in heat source temperature control over conventional thermal control pipes can be achieved.

(iii)   The greatest benefits of feedback control are realised where high source resistances (R_s) are present or large variations in heat load or environmental conditions occur.

(iv)   Active feedback control gives better temperature control than the equivalent passive system and also mass diffusion and varying sink conditions are less detrimental in an active system.

The model resulting from this study describes in differential form the variation in source temperature affected by changes in load, sink conditions and
other independent variables. The model is quite general in that it applies
to no specific configuration and treats the methods of obtaining control in
functional form. The only limitation imposed upon the system is that a non-
condensing gas is used to control the available heat rejection area.

The basic assumptions of the analysis are that:

    (i)   The non-condensing gas obeys the ideal gas law.

    (ii)   Steady state conditions exist.

    (iii) A sharp interface exists between the gas and the vapour.

(In cases where axial heat conduction is appreciable the latter assumption is
inconsistant with experimental observations which show that the decrease in
vapour concentration and the corresponding increase in gas concentration
occur smoothly over an appreciable length of the heat pipe. However, if
axial conduction is minimised this assumption is reasonable and as will be
described later the work of Edwards and Marcus (6.17) allows the effect of
axial heat conduction and mass diffusion to be taken into account).

Simultaneous solution of the equations of energy and mass conservation, state
of the non-condensable and auxiliary fluid (passive system), force balance
and auxiliary functions relating to variation in source temperature
(eqn. 6.1), the total pressure and pressure/temperature relationships,
results in eqn. 6.3.

$$d\,T_s = \frac{\{\dfrac{\partial T_v}{\partial Q} + R_s\,(1 + S + S_1)\}\,dQ + \{\dfrac{\partial T_v}{\partial T_0} + S\,\psi_0\}\,d\,T_0 + S\,\psi_{ST}\,d\,T_{ST}}{1 + S + S_1 + S_2}$$

$$\ldots\,6.3$$

This equation describes the change in source temperature (T_s) as a function
of changes in heat load (Q), sink temperature (T_0), gas storage temperature
(T_{ST}). This equation is related to the vapour temperature through eqn. 6.1
and is particularly useful for the examination of the effect of the control
parameter S, S_1 and S_2 which are expressed in functional form. The function

ψ (subscript 0 - sink conditions, ST - storage condition) is associated with establishing conservation of the mass of non-condensable within the inactive part of the condenser and the storage volume i.e. it relates to the influence of sink or storage temperature on the displacement of the non-condensable gas.

$$\psi = \frac{1}{(p\alpha)_v} \left[P_{vis} \, \gamma \, (1 + \frac{\partial \, P_v}{\partial P_{vis}}) + \frac{\partial P_g}{\partial T} \right] \qquad \ldots \, 6.4$$

where P_v is the vapour pressure

        P_{vis} is the vapour pressure in the inactive condenser region or the storage region

and α_v = $\partial \ln P_v \, / \partial T_v$

       γ = $\partial \ln P_{vis} \, / \partial T$

       S   is a control parameter relating to the use of a non-condensable gas and a fixed storage volume and is present for both passive and active control.

       S_1 applies when a variable storage volume is used but actuated by internal vapour pressure changes (S_1 = 0 for active control).

       S_2 applies when passive feedback control is used such as an auxiliary fluid sensing the source temperature (S_2 = 0 for active control).

Examination of eqn. 6.3 clearly shows the effect of each variable and also indicates the required magnitudes of the control parameters to provide good attenuation of variation in source temperature and stability of operation. For example, with the passive feedback system if the heat load is the main variable i.e. sink and storage temperature constant ($\partial T_0 = 0$, $\partial T_{ST} = 0$), then S_2 must be large compared to $(1 + S + S_1)$ if good attenuation is to be achieved. This immediately indicates that the auxiliary fluid should be a liquid since this gives much greater values of S_2, and that S_1 should be small or negative. However, if the sink temperature varies it is desirable to have $S_1 > 0$ and $(S_1 + S_2)$ large compared to S. Optimisation of the passive system is difficult in the presence of multi-variations but for $S_2 > 0$ improved control compared to the conventional system must be achieved.

The disadvantages of the passive system indicate that the active system will be the most desirable design and henceforth the discussion will be orientated toward that system i.e. $S_1 = S_2 = 0$. Thus eqn. 6.3, immediately reduces to

$$d\,T_s = \frac{\{\frac{\partial\,T_v}{\partial\,Q} + R_s\,(1 + S)\}\,d\,\theta + \{\frac{\partial\,T_v}{\partial\,T_0} + S\,\psi_0\}\,d\,T_0 + S\,\psi_{ST}\,d\,T_{ST}}{1 + S}$$

$$\ldots\ 6.5$$

In the active system control is effected by changing the saturation temperature of the working fluid in the storage volume by means of an auxiliary heater. This is an active way of essentially varying the volume available for the non-condensable. Thus control is achieved by the magnitude of change of dT_{ST} and ψ_{ST} in eqn. 6.5. This indicates that S should be sufficiently large to render $\frac{S}{1 + S} \simeq 1$.

6.6.2 Storage volume requirements for active feedback control. The storage volume requirements for active feedback control are determined from mass conservation considerations, i.e.

$$M_g = \frac{(P_v - P_{vi})\,V_{ic}}{R_g\,T_0} + \frac{(P_v - P_{vST})\,V_{ST}}{R_g\,T_{ST}} \qquad \ldots\ 6.6$$

where M_g is the mass of non-condensable

P_v is the vapour pressure

P_{v0} is the vapour pressure in the inactive part of the condenser (i.e. at T_0).

V_{ic} is the volume of the inactive part of the condenser

V_{ST} is the storage volume

P_{vST} is the vapour pressure in the storage volume

T_O is the sink temperature

T_{ST} is the storage temperature

R_g is the characteristic gas constant (non-condensable)

The storage requirement may now be determined from the examination of the heat pipe operating extremes:

(i)   At the high power condition all of the non-condensable gas should be contained within the storage volume. This implies that the storage volume should be at the lowest temperature it can achieve (i.e. the maximum sink temperature) and also that a working fluid of low vapour pressure at this temperature be used.
Thus

$$Mg = \frac{(P_v - P_{vST})_H \, V_{ST}}{R_g \, T_{OH}} \qquad \cdots 6.7$$

(subscript H indicates high power conditions)

(ii)   At the lower power condition all of the non-condensable should be in the condenser sections. This implies that the partial pressure of the vapour in the storage volume approaches the system vapour pressure. However, from a practical consideration, in order to minimise mass diffusion and the resulting poor temperature control, the storage temperature should be somewhat lower than the heat pipe vapour temperature. This allows some concentration of non-condensable in the storage volume and hence reduces the potential for mass diffusion. Acceptable storage temperatures at the low power condition must be determined from the diffusion characteristics for the system. This was beyond the scope of the investigation (6.6, 6.7), but can be investigated using the model of Edwards and Marcus (6.17).

Thus for the low power condition:

$$M_g = (\frac{P_v - P_{vi}}{R_g T_0})_L V_{ci} + (\frac{P_v - P_{vST}}{R_g T_{ST}})_L V_{ST} \qquad \ldots 6.8$$

If the storage temperature is set equal to the vapour temperature the last term in eqn. 6.8 is zero. Simultaneous solution of eqns. 6.7 and 6.8 gives the storage volume requirements

$$\frac{V_{ST}}{V_{ci}} = \frac{(1 - \frac{P_{vi}}{P_v})_L (\frac{T_{ST}}{T_0})_L}{(\frac{P_{vH}}{P_{vL}}) (1 - \frac{P_{vST}}{P_v})_H (\frac{T_L}{T_H})_{ST} - (1 - \frac{P_{vST}}{P_v})_L} \qquad \ldots 6.9$$

The storage volume is related to the variation in heat load by eqn. 6.1. Hence, the required variation in vapour temperature can be calculated if the source/vapour heat transfer resistance can be estimated. One point which is immediately apparent from eqn. 6.9 is that storage requirements will be larger for a working fluid which has a low vapour pressure at the nominal operating condition since, in general,

$$\frac{(P_v - P_{vi})_L}{(P_v - P_{vST})_H}$$

will be larger for a low pressure fluid. Thus, to minimise storage requirements (and hence auxiliary power requirements) a fluid with a high vapour pressure at the operating condition should be selected. However, this is inconsistant with the existance of rapid transient response and an optimum thermal control system can be designed only when transient behaviour of the source temperature has been considered.

Once V_{ST} has been determined eqn 6.7 may be solved for the required mass of non-condensable.

6.6.3 Transient analysis.

The worst transient condition that can be experienced by a heat pipe is a step increase or decrease in power input at the evaporator.

The transient analysis for this condition has been studied by Bienert and

Brennan (6.6).

Since the governing equations for variable conductance heat pipes are highly
non-linear and large changes in system variables are associated with step
changes, described above, the following simplifying assumptions are made.

(i)   The mode of heat dissipation can be approximated by the
       convection equation i.e. fourth order terms associated
       with radiation are eliminated.

(ii)   Recovery of the vapour temperature in the heat pipe proper
       occurs at the same rate as that of the storage volume.

$$\text{i.e. } \left(\frac{T - T_H}{T_{in} - T_H}\right)_V = \left(\frac{T - T_H}{T_L - T_H}\right)_{ST}$$

       where T is the instantaneous temperature

               T_H is the high power temperature

               T_L is the low power storage temperature

               T_{in} is the initial vapour temperature

       This implies that the vapour temperature responds instant-
       aneously to changes in storage temperature. This is valid
       if the resistance between the storage reservoir and sink
       is high compared to that between the vapour and sink.
       This is reasonable, as the storage volume will normally
       be insulated to minimise auxiliary power requirements.
       This means that the time constant for the condenser
       ($\tau = M\ C_p\ R$) is small compared to that of the reservoir,
       i.e. $\dfrac{\tau_c}{\tau_{ST}} \ll 1$.

(iii) An ideal on/off controller is used.
       The heat input to the storage volume for a low to high
       power condition may now be written as

$$Q_{ST} = (M\ C_p)_{ST}\ \frac{d\ T_{ST}}{dt} + (hA)_{ST}\ (T_{ST} - T_{OH}) \qquad \dots\ 6.10$$

where Q_{ST} is the heat input to the storage volume

         M_{ST} is the mass of the storage volume

         C_p is the specific heat of the storage volume

         T_{ST} is the instantaneous storage temperature

         T_{OH} is the effective sink temperature at high
                         power conditions

         $(hA)_{ST}$ is the storage/sink conductance.

The maximum auxiliary power requirement will occur at the low power condition at which point the storage temperature will be approaching the system vapour temperature. This condition governs the insulation requirements of the storage volume.

i.e. $(hA)_{ST}$ $=$ $\dfrac{Q_{ST}}{(T_v - T_0)_L}$ ... 6.11

where T_v and T_0 are the vapour and sink temperatures at the low power condition.

Solving eqns 6.10 and 6.11 gives the recovery time of the storage temperature and hence vapour temperature

$$\left(\frac{T - T_H}{T_{in} - T_H}\right) = e^{\dfrac{-t}{\tau_{ST}}}$$ ... 6.12

where τ_{ST} $=$ $\dfrac{(M\,C_p)_{ST}\,(T_v - T_0)_L}{Q_{ST}}$ ... 6.13

Thus the storage and vapour temperature will vary exponentially in response to a power change. In practice a slight overshoot/undershoot of vapour temperature will occur before the final steady state value is obtained but this will be slight if the reservoir response is rapid.

At the heat source

$$(Q_S)_H = (M\,Cp)_S\,\frac{d\,T_S}{d\,t} + \frac{T_S - T_v}{R_S}$$ ... 6.14

Substitution of eqn. 6.1 and 6.12 into 6.14 and integrating gives the response time of the source to step increase in power

$$\psi = \frac{(T - T_n)_S}{(T_{in} - T_H)_v} = \frac{1}{1 - \dfrac{\tau_S}{\tau_{ST}}} (e^{-\frac{t}{\tau_{ST}}} - e^{-\frac{t}{\tau_S}}) \qquad \ldots \, 6.15$$

where $\tau_S = (M\,Cp\,R)_S$ $\qquad\qquad\qquad \ldots \, 6.16$

$(T_n)_S$ is the nominal source operating temperature

T is the instantaneous source temperature

T_{in}, T_H are the initial and final vapour temperatures.

Eqn. 6.15 also applies to a step decrease in power if the maximum auxiliary power is just sufficient to maintain equilibrium conditions at the low power conditions.

The maximum overshoot/undershoot at the source is obtained by differentiation of eqn. 6.15.

$$\psi_p = \frac{(T_p - T_n)_S}{(T_{in} - T_H)_v} = (\frac{\tau_{ST}}{\tau_S})^{\dfrac{1}{\frac{\tau_{ST}}{1 - \frac{\tau_{ST}}{\tau_S}}}} \qquad \ldots \, 6.17$$

and the corresponding time

$$t_p = \frac{\tau_S}{1 - \dfrac{\tau_S}{\tau_{ST}}} \, \ell n \, \frac{\tau_{ST}}{\tau_S} \qquad\qquad \ldots \, 6.18$$

where ψ_p is the maximum overshoot/undershoot temperature and other parameters as defined earlier.

The following comments can now be made for a step increase in power:

(i)   The source temperature increases exponentially up to the point of maximum overshoot after which it decreases exponentially to become asymptotic to its nominal value.

(ii)   The value of $\dfrac{\tau_{ST}}{\tau_S}$ (eqn. 6.17) should be as small as possible
to reduce the maximum overshoot and minimise recovery time
(eqn. 6.18). The most efficient way of doing this is to
minimise the heat capacity of the reservoir. A reduction
in the reservoir's insulation will also help improve
response to increases in heat load and/or sink temperature
but will result in an increased auxiliary power requirement.

The maximum overshoot/undershoot and the time to achieve this condition is
shown in Fig. 6.7. This shows quite clearly the necessity for a small
$(\dfrac{\tau_{ST}}{\tau_S})$.

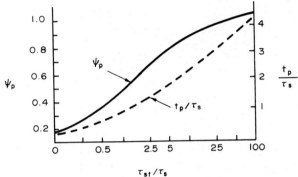

Fig. 6.7   Transient response of a VCHP

The recovery time (t_r), defined as the time required for the controlled
variable (T_S) to come within a specified absolute percentage of its final
value and thereafter remain less than the specified percentage is shown in
Fig. 6.8, for different values of percentage (ψ_r). This shows that the
recovery time begins to increase rapidly if $\dfrac{\tau_{ST}}{\tau_S}$ exceeds unity, i.e. the time
constant of the reservoir exceeds that of the source.

Bienert and Brennan used this theory to check the performance of a water/
argon electrical feedback VCHP and in general obtained good agreement.

The theory predicts a shorter time to the overshoot and a lower value of
overshoot of source temperature because of the assumption that the system
vapour temperature and hence pressure respond instantaneously to change in
power and/or sink conditions and that the vapour temperature itself does not

experience any overshoot/undershoot before it begins to recover. The magnitude of the vapour overshoot depends upon the response of the reservoir relative to the source i.e. the theory will be more accurate for small values of $\dfrac{\tau_{ST}}{\tau_S}$ (in Bienert and Brennan's case $\dfrac{\tau_{ST}}{\tau_S}$ was approximately 44).

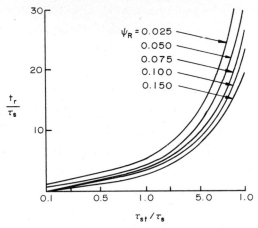

Fig. 6.8 Transient characteristics - recovery time

The theory also overpredicts the recovery time since it is based upon the storage temperature asymptotically approaching its final value. In practice the vapour and source temperatures approach their nominal value before the storage temperature reaches its equilibrium value, because near to extreme points, small deviations of storage temperature have little effect on vapour or source temperature.

6.7 Comparison of Theory and Experiment - Active Feedback Control

The transient performance obtainable in practice on VCHP's may be illustrated using results obtained on an electrical feedback controlled water/argon system developed at IRD, and illustrated in Fig. 6.9. The heat pipe had the following specification:

Source description	4 resistances
Source temperature	$70^{\circ}C$
Minimum sink temperature	$10^{\circ}C$
Maximum sink temperature	$30^{\circ}C$
Giving $\dfrac{V_{ST}}{V_c}$	$= 2.75$

Evaporator length	150 mm
Condenser length	300 mm
Heat pipe bore	12.5 mm
Storage bore	25 mm
Storage length	160 mm
Wick	4 layers 200 mesh
Working fluid	Water
Control gas	Argon
Maximum power	100 W
Mass of gas	233×10^{-6} kg
Auxiliary power	15 W

Fig. 6.9 Variable conductance heat pipe

Following stabilisation at the low power/low sink condition, the power input at the evaporator was increased from 10W to 78 W and the sink temperature to 30°C. The result was as shown in Fig. 6.10.

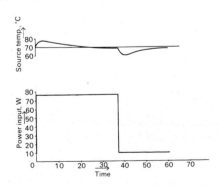

Fig. 6.10 Experimental and theoretical transient performance of the IRD VCHP.

After an initial temperature overshoot, the source returned to within $1^{\circ}C$
of the set value within 20 minutes. Similarly a step reduction in power
input was followed by an undershoot of $9^{\circ}C$, but the source temperature was
only $1^{\circ}C$ below the set point after 20 minutes.

Comparing the experimental results with those predicted by the theory
presented above, the time to peak t_p given by eqn. 6.18 was 136 sec, and was
measured to be 150 sec. The measured peak overshoot and undershoot exceeded
that calculated, owing to the fact that the theory assumes instantaneous
vapour temperature and pressure response to changes in power. The magnitude
of the overshoot (and undershoot) is a function of the storage volume response
response.

6.8 Effects of the Non-condensable Gas on the Working Fluid in Variable Conductance Heat Pipes

Most theories for variable conductance heat pipes are based on the assumption
that the gas/vapour interface is sharp and diffusion between the two regions
is not present. In practice this is not the case, and in some designs it is
necessary to take diffusion into account.

A second more serious phenomenon resulting from the introduction of inert gas
control into a heat pipe occurs when gas bubbles enter the wick structure,
via the working fluid.

These two features are discussed below.

6.8.1 Diffusion at the vapour/gas interface.
Several investigators have
questioned the assumption that a sharp vapour/gas interface exists in gas-
loaded heat pipes (6.3, 6.17). It has been demonstrated that in some gas-
buffered heat pipes the energy and mass diffusion between the vapour and non-
condensable gas could have an appreciable effect on heat transfer in the
interface region and the temperature distribution along the heat pipe.

Two dimensional analyses of the interface region are reported by Somogyi and
Yen (6.18) and Rohani and Tien (6.19) and may be used to investigate this
phenomenon. The analyses showed that lowering of the sink temperature had
the effect of reducing the width of the interface diffuse region, which can

be 3 - 4 cm in length. The diffusion coefficient of the inert gas has an
effect on the extent of the diffuse region, gases having higher diffusion
coefficients being less desirable, reducing the maximum heat transport
capability of the heat pipe by reducing local condenser temperature. It must
be noted that the diffusion coefficient is inversely proportional to density,
and therefore at lower operating temperatures, particularly during start-up
of the heat pipe, the diffuse region may be extensive and of even greater
significance. It is therefore important to cater for this during any
transient performance analysis.

6.8.2 Gas bubbles in arterial wick structures.

Although in simple heat
pipes containing only the working fluid, freeze-degassing of the liquid (see
Chapter 4) can remove any dissolved gases; in a variable conductance heat
pipe inert gas is always present. If the gas dissolves in the working fluid,
or finds its way in bubble form into arteries carrying liquid, the performance of the heat pipe can be adversely affected.

Saaski (6.20) has carried out theoretical and experimental work on the isothermal dissolution of gas in arterial heat pipes, examining effects of
solubility and diffusivity of helium and Argon in ammonia, Freon 21 and
methanol.

One of the significant factors determined by Saaski was the venting time
of bubbles in working fluids (the time for a bubble to disappear).

The venting time t_v may be calculated from the equation

$$t_v = \frac{R_0^2}{3 \alpha D}$$

where R_0 is the bubble radius (initial)

α is the Ostwald coefficient, given by the ratio
of the solute concentration in the liquid phase
to the concentration in the gaseous phase (6.21)

D is the diffusion coefficient.

Predicted values of t_v are given in Table 6.2.

TABLE 6.2   VENTING TIME OF GAS BUBBLES IN WORKING FLUIDS

(R_0 = 0.05 cm)

Fluid	Temperature	t_v(sec)	
		Helium	Argon
Ammonia	-40	1200	107
	20	63	6.7
	60	7	1.6
Freon 21	-40	367	43
	20	67	17.5
	60	23	7.5
Methanol	-40	1030	154
	20	133	55
	60	50	26
Water	22	1481	1215

Table 6.2 shows that venting times can be considerable when the working fluid is at a low temperature, but in general argon is more easily vented than helium.

The equation above is not valid when non-condensable gas pressure is significant compared to the value of 2 σ_ℓ/R_0, where σ_ℓ is the surface tension of the working fluid. Saaski stated that venting time increases linearly with non-condensable gas pressure, other factors being equal, and showed that if, as in a typical gas-controlled heat pipe, the helium pressure is about equal to the ammonia working fluid vapour pressure, the vent time can be 9 days. This is a very long time when compared with the transients to be expected in a VCHP; by changing the working fluid and/or control gas, relatively long venting times may still be obtained.

Having established venting times for spherical bubbles, Saaski developed a theory to cater for elongated bubbles, the type most likely to form in arteries. He obtained the results in Table 6.3 for the half lives of elongated arterial bubbles in a VCHP at 20°C, (artery radius 0.05 cm, non-condensable gas partial pressure equal to vapour pressure).

The models used to calculate these values were confirmed experimentally, and it was concluded that the venting times are of sufficient length that repriming of an arterial heat pipe containing gas (see Chapter 3) may be

TABLE 6.3   HALF LIVES OF ARTERIAL BUBBLES IN VARIOUS
WORKING FLUIDS

Fluid	$t_{\frac{1}{2}}$ (helium)	$t_{\frac{1}{2}}$ (argon)
Ammonia	7 days	17 hrs
Freon 21	1.5 days	9.5 hrs
Methanol	4.8 hrs	1.7 hrs
Water	3 hrs	2.5 hrs

possible only if some assistance in releasing gas occlusions can be given
during start-up or steady state operation, either by internal phenomena or
external interference.

Kosson et al (6.2) introduced another factor affecting VCHP's, namely variations in pressure within the pipe due to oscillations in the diffusion zone.
These pressure variations are of the same order as the capillary pressure and
can cause vapour flashing within the artery, with accompanying displacement
of liquid from the artery.

In order to overcome this and occlusion problems, subcooling of the liquid in
the artery was carried out by routing the fluid to the condenser wall so that
it experienced sink conditions before returning to the evaporator. As shown
in Saaski's results, lowering the liquid temperature improved venting time.
It was also found to reduce the sensitivity of the artery to vapour formation
caused by the pressure oscillations described above.

Thus, while inert gases can create problems in VCHP's, sufficient data is
available to enable the designer to minimise its detrimental effects.

6.9 VCHP Computer Programme

A computer programme is available (6.22) which is very useful in the design
and analysis of heat pipes containing non-condensable gases, either for temperature control or to aid start-up from a frozen state. The programme can
cater for diffusion, as discussed in Section 6.8, and also allows for
conduction along the heat pipe wall in the region of the gas/vapour interface.

The programme is listed in the referenced report, which also includes worked

HP—I

examples and details of data input. It is not intended to present this here,
but the programme allows one to carry out the following calculations:

(i)   Calculate the wall temperature profile along a gas-buffered
        heat pipe.

(ii)   Determine the gas inventory needed to obtain a desired evap-
        orator temperature corresponding to the required heat load.

(iii)   Calculate duty/evaporator temperature characteristic for a
        fixed amount of gas in the heat pipe.

(iv)   Calculated the heat and mass transfer along the pipe, including
        the diffuse front region.

(v)   Calculate the heat leakage rate for the case when the condenser
        is full of gas.

(vi)   Predict freezing rate in the condenser, if it occurs.

(vii)   Determine information necessary to size the reservoir.

The programme may be applied to hot or cold reservoir passive VCHP's, as well
as active feedback-controlled systems. Other options available cover most of
the situations arising during off-design performance and overall system geometry.

The manual assumes that the user has a prior knowledge of gas-buffered heat
pipe principles, as given in this chapter and the numerous references cited.

6.10 Recent Developments

The significance of variable conductance heat pipes is evident from the fact
that almost 15 per cent of the papers at the 1976 International Heat Pipe Conference dealt directly with this topic, and many others contained reference to
their performance and application. Groll and Kirkpatrick (6.23) reviewed the
state of the art in this technology, concentrating on spacecraft applications
which have set the pace for developments in this area. Their review is best
summarized with reference to Table 6.4. Several of the VCHP types presented
here are discussed earlier, where we concentrated primarily on gas-buffered
heat pipes. Control of the vapour flow by throttling, and liquid control (see
Section 5.6) are receiving increasing attention.

Groll and Kirkpatrick pointed out the potential of electro-osmotic and EHD heat pipes for temperature control. As the capability of these types of heat pipe is a function of the applied voltage, variations in the voltage can be used to effect control, and the addition of a feedback system appears feasible.

TABLE 6.4.   Comparison and evaluation of heat pipe control techniques

CONTROL TECHNIQUE	CRITERION FOR COMPARISON	HEAT TRANSP CAPA BILITY	VOLUME/ MASS REQUIRE MENTS	COMPLEX ITY EASE OF FABRIC	RELIA BILITY	CONTROL CHARACTERISTICS	STATE OF THE ART/ SPACECRAFT	
VARIABLE CONDUCTANCE TECHNIQUE	NON-FEEDBACK					ONLY VAPOR TEMPERATURE CAN BE CONTROLLED		
	(i) WICKED (COLD) GAS RESERVOIR	F to G[1] VG[2]	F	G	G	A[4]: PASSIVE; SIMPLE DESIGN; DIODE POTENTIAL. D[5]: VERY SENSITIVE TO SINK TEMP. VARIATIONS, ESPECIALLY WHEN SINK TEMP IS CLOSE TO OPERATING TEMP. (CIRCUMVENT BY USE OF CONSTANT TEMP. RESERVOIR); ARTERIAL WICKS PROBLEMATIC.	4	CTS
	(ii) NON-WICKED (HOT) GAS RESERVOIR	F to G[1] VG[2]	F	G	F to G	A: PASSIVE; SIMPLE DESIGN; LESS SENSITIVE TO SINK TEMP VARIATIONS. D: SENSITIVE TO PRESENCE OF VAPOR OR LIQUID IN RESERVOIR (MINIMIZE WITH VERY SMALL OR VERY LARGE L/D GAS FEED TUBES); ARTERIAL WICKS PROBLEMATIC.	4	OAO-3
	(iii) CASCADED VCHP (NON WICKED GAS RESERVOIR)	F to G[1] VG[2]	F[6]	F to G	F to G	A: PASSIVE; EVEN LESS SENSITIVE TO SINK TEMP. VARIATIONS (IF COARSE CONTROL PIPE HAS NON WICKED RESERVOIR). D: TWO HEAT PIPES NEEDED; LARGER TOTAL TEMP. DROP; ARTERIAL WICKS PROBLEMATIC.	2-3	
	FEEDBACK					SOURCE TEMPERATURE CAN BE CONTROLLED		
	(i) ELECTRICALLY HEATED WICKED GAS RESERVOIR	F to G[1] VG[2]	F to G	F to G	G	A: VERY GOOD CONTROL WITH SMALL RESERVOIRS; GOOD TRANSIENT RESPONSE; VARIABLE SET POINT POSSIBLE; DIODE POTENTIAL. D: ELECTRICAL POWER AND CONTROLLER REQUIRED; ARTERIAL WICKS PROBLEMATIC.	4	ATS-6
	(ii) BELLOWS SYSTEM ACTIVATED BY EXPANSION FLUID, NON-WICKED GAS RESERVOIR	F to G[1] VG[2]	P	P to F	F	A: GOOD CONTROL; NO ELECTRICAL POWER REQUIRED. D: MECHANICALLY COMPLEX, LARGE AND HEAVY, ARTERIAL WICKS PROBLEMATIC.	2	
	(iii) VMHP[3]: VAPOR FLOW RATE CONTROL	P to F	G	F	F	A: GOOD CONTROL; NO ELECTRICAL POWER REQUIRED; INSENSITIVE TO SINK TEMP. VARIATIONS. D: LIMITED TO LOW PRESSURE FLUIDS AND FINE CAPILLARY STRUCTURES (BLOW THROUGH LIMIT); FREEZING PT. OF FLUID MUST BE BELOW SINK TEMP.	2	
	(iv) VMHP[3]: INDUCED WICK/GROOVE DRY OUT TECHNIQUE	F[1] to VG[2]	G	F	F to G	A: SAME AS ABOVE, BUT NO RESTRICTIONS ON FLUID OR WICK DUE TO BLOW THROUGH LIMIT. D: COMPLEX (BUT COMPACT); LARGER OVERALL TEMP. DROP IN 2 AND 3 HEAT PIPE CONFIGURATIONS.	2	
DIODE TECHNIQUE	(i) EVAPORATOR BLOCKAGE BY LIQUID	G[2]	G	G	G	A: PASSIVE; SMALL LIQUID RESERVOIR (IF EVAPORATOR AND VAPOR SPACE ARE SMALL); GOOD TRANSIENT RESPONSE. D: 1 g TESTING DIFFICULT AND VAPOR PRESSURE DROPS CAN BE HIGH (IF BLOCKING ORIFICE IS NOT USED).	4	ATS-6
	(ii) EVAPORATOR BLOCKAGE BY GAS	F to G[1] VG[2]	F to G	G	G	A: PASSIVE; SIMPLE DESIGN; CAN BE USED WITH AXIAL GROOVES; CAN BE USED AS VCHP. D: LARGER RESERVOIR AND POORER RESPONSE THAN LIQUID BLOCKAGE; ARTERIAL WICK PROBLEMATIC.	2-3	
	(iii) TRAPPING OF WORKING FLUID	F to G[1] VG[2]	G	G	G	A: PASSIVE; SMALL TRAP AND GOOD TRANSIENT BEHAVIOR (DEPENDS ON WICK); CAN BE USED WITH GAS VCHP IN HYBRID SYSTEM; CAN BE USED AS THERMAL SWITCH. D: LARGE TRAPS REQUIRED BY ARTERIAL WICKS; ENERGY MUST BE DISSIPATED FROM TRAP DURING SHUTDOWN.	2	
HYBRID TECHNIQUE	GAS CONTROL VCHP AND GAS BLOCKAGE DIODE	F to G[1] VG[2]	F	G	G	A: PASSIVE; ONE PIPE SERVES TWO PURPOSES; SIMPLE IF AXIAL GROOVES ARE USED; MOST VCHP TECHNIQUES APPLICABLE. D: ARTERIAL WICKS MAY BE PROBLEMATIC.	1	

CLASSIFICATIONS

P = POOR   1 = RESEARCH
F = FAIR   2 = LABORATORY PROTOTYPE
G = GOOD   3 = FLIGHT HARDWARE
VG = VERY GOOD   4 = FLIGHT DEMONSTRATION

1) FOR NON ARTERIAL HEAT PIPES
2) FOR ARTERIAL HEAT PIPES
3) VMHP = VAPOR FLOW MODULATED HEAT PIPE
4) A = ADVANTAGES
5) D = DISADVANTAGES
6) HOWEVER, 2 OR MORE VCHP's USED

Practical experience of VCHP's in satellite applications, notably on OAO-3 (the Ames Heat Pipe Experiment - AHPE) and on ATS-6, has shown that the heat pipes have performed well to date, OAO-3 being launched in August 1972 and ATS-6 entering orbit in May 1974. A further three units are likely to be launched within the next few years.

Terrestrial applications of sophisticated gas controlled heat pipes are grow-

ing, and work described by Kelleher and Batts (6.24) draws attention to possible difficulties arising in their operation. On tests with heat pipes using
methanol as the working fluid and either krypton or helium as the control gas,
it was found that if the working and control gas possessed significantly differing molecular weights, orientation in a gravitational field could seriously
affect operation. On small diameter heat pipes, (a 1.6 cm diameter unit being
tested), the effect was to distort the vapour/gas interface and to displace it,
without seriously interfering with the one-dimensional nature of the transport
in the pipe. Tests on a 5 cm diameter heat pipe revealed stratification of the
vapour and gas.

The authors did not take into account other operating limitations in the heat
pipes (for example in a 5 cm diameter heat pipe operating horizontally, circumferential wicking limits may be encountered) but their work suggests that care
should be taken when using similar working fluid/gas mixes in VCHP's.

Other work on variable conductance heat pipes is of a longer term nature, but
several interesting techniques for implementing control were discussed. These
included a soluble gas absorption reservoir, in which a liquid matrix or sponge
replaces the normal gas reservoir, (6.25). Advantages claimed include much reduced reservoir volume for a required degree of temperature control. Vapour
flow control methods were discussed by Marcus and Eninger (6.26), and the use
of the vapour lift pump, also applied in the 'anti-gravity thermosyphon' described in Chapter 5, has been successfully applied at Bell Telephone Laboratories (6.27). As discussed in Chapter 5, the operation of this system necessitates the application of heat to the bubble pump as well as to the evaporator
section of the heat pipe. However, if the system is compared with the feedback
controlled VCHP, additional electrical loads are unlikely to be any higher.

The transverse header.

Entrapment of gas bubbles in arteries in VCHP's has been described in Section
6.8.2. This problem can be overcome using a VCHP developed by NASA, which involves rearranging the evaporator and condensers surfaces to make them insensi-
tive to bubble formation. Such a change results in short liquid transport
lengths, enabling high heat transport capabilities to be achieved.

The transverse header, illustrated in Fig. 6.11, consists of a sealed volume
of rectangular cross section, (6.28).

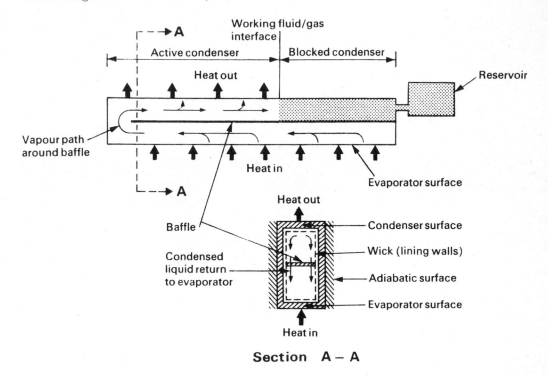

Fig 6.11 Transverse header variable conductance
          heat pipe (6.28)

The bottom and top surfaces serve as the evaporator and the condenser respectively, while the side walls are adiabatic surfaces. The inside walls are wicked along their entire length with either screen mesh or circumferential (rather
than axial) grooves, and a thin solid baffle is positioned between the evaporator and condenser surfaces dividing the vapour space into two. The baffle
runs the length of the unit, and except for the opening at the left, is sealed
against the walls to prevent vapour 'short circuiting' directly to the condenser surface. The seal is not so tight, however, as to prevent condensed liquid
from being wicked directly from the condenser to the evaporator down the side
walls. This allows the liquid, whose pressure losses are usually high in conventional heat pipes, to take the shortest path from condenser to evaporator;
the vapour's losses are usually low and it travels the longer route.

Temperature control is provided by the usual noncondensable gas reservoir which communicates with the heat pipe through the condenser vapour space between the baffle and condenser surface. Blockage of the condenser vapour proceeds from right to left, with complete condenser blockage occurring when the fluid/gas interface travels across the entire condenser length. In this condition the residual losses consist of conduction along both wick and side walls, as well as vapour diffusion through the blocked condenser vapour space. These losses tend to be higher in the transverse header VCHP compared with the conventional VCHP because of the closeness of evaporator and condenser. When the transverse header is operated below maximum load the fluid/gas interface automatically adjusts to changing environments and load in an attempt to maintain constant vapour temperature. A significant benefit of the new device is that movement of the gas across the wick does not impair wick performance.

Although a device like that illustrated has been built and tested, a different physical configuration of the transverse header was used as a prototype for spacecraft applications. Loads as high as 3.6 kW have been obtained with this prototype- higher than that attained with conventional VCHP's (See also Appendix A7).

The variable conductance heat pipe is probably the most significant development in heat pipe technology. Further information on uses is given in Chapter 7.

REFERENCES

6.1   Wyatt, T. A controllable heat pipe experiment for the SE-4
        satellite. JHU Tech. Memo. APL-SDO-1134, Johns Hopkins University,
        Applied Physics Lab., AD 695433, March 1965.

6.2   Kosson, R. et al. Development of a high capacity variable conduc-
        tance heat pipe. AIAA Paper 73-728, 1973.

6.3   Marcus, B.D. and Fleischman, G.L. Steady state and transient per-
        formance of hot reservoir gas controlled heat pipes. ASME Paper
        70-HT/SpT-11, 1970.

6.4   Turner, R.C. The constant temperature heat pipe - a unique device
        for thermal control of spacecraft components. AIAA 4th Thermo-
        physics Conf., Paper 69-632, June 1969 (RCA).

6.5   Rogovin, J. and Swerdling, B. Heat pipe applications to space
        vehicles. AIAA Paper 71-421, 1971.

6.6   Bienert, W. and Brennan, P.J. Transient performance of electrical
        feedback-controlled variable conductance heat pipes. ASME Paper
        71-Av-27, 1971.

6.7   Bienert, W., Brennan, P.J. and Kirkpatrick, J.P. Feedback-con-
        trolled variable conductance heat pipes. AIAA Paper 71-421, 1971.

6.8   Bienert, W. et al. Study to evaluate the feasibility of a feed-
        back-controlled variable conductance heat pipe. Contract NAS2-
        5772, Tech. Summary Report DTM-70-4, Dynatherm, Sept. 1970.

6.9   Depew, C.A., Sauerbrey, W.J. and Benson, B.A. Construction and
        testing of a gas-loaded passive control variable conductance heat
        pipe. AIAA Paper 73-727, 1973.

6.10   Bienert, W. Heat pipes for temperature control. AIChE 4th Inter-
        soc, Energy Conf. Washington, D.C., Sept. 1969.

6.11   Katzoff, S. Heat pipes and vapour chambers for the thermal control
        of spacecraft. AIAA Paper 67-310, April, 1967.

6.12   Kirkpatrick, J.P. and Marcus, B.D. A variable conductance heat
        pipe experiment. AIAA Paper 71-411, 1971.

6.13   Anand, D.K. and Hester, R.B. Heat pipe applications for space-
        craft thermal control. Johns Hopkins University. Tech. Memo.
        TG-922, Aug. 1967.

6.14   Schlosinger, A.P. Heat pipes for space suit temperature control.
        Aviation and Space Conference, AIAA, Los Angeles, June, 1968.

6.15   Katzoff, S. Notes on heat pipes and vapour chambers and their
        applications to thermal control of spacecraft. USAEC Report
        SC-M-66-623. Proc. Joint Atomic Energy Commission/Sandia Lab.
        Heat Pipe Conf., Vol. 1, pp 69 - 89, June, 1966.

6.16   Marcus, B.D. Theory and design of variable conductance heat pipes.
        NASA CR-2018, April, 1972.

6.17   Edwards, D.K. and Marcus, B.D. Heat and mass transfer in the
        vicinity of the vapour/gas front in a gas loaded heat pipe.
        Submitted for presentation at ASME Winter Annual Meeting, Washing-
        ton, D.C., Nov. 1971, Pub. Trans. ASME, Ser.C, Vol. 94, No.2, 1972.

6.18   Somogyi, D. and Yen, H.H. An approximate analysis of the
       diffusing flow in a self-controlled heat pipe. ASME J. Heat
       Transfer, Feb. 1973.

6.19   Rohani, A.R. and Tien, C.L. Steady two-dimensional heat and mass
       transfer in the vapour gas region of a gas-loaded heat pipe. ASME
       J. Heat Transfer, Aug. 1973.

6.20   Saaski, E.W. Gas occlusions in arterial heat pipes. AIAA Paper
       73-724, 1973.

6.21   Saaski, E.W. Investigation of bubbles in arterial heat pipes.
       NASA CR-114531, 1973.

6.22   Edwards, D.K., Fleischman, G.L. and Marcus, B.D. User's manual
       for the TRW gas pipe 2 program. NASA CR-114672, TRW Systems Group
       Oct. 1973.

6.23   Groll, M. and Kirkpatrick, J.P. Heat pipes for spacecraft tempera-
       ture control - an assessment of the state of the art. Proc. 2nd
       Int. Heat Pipe Conference, Bologna; ESA Report SP 112, 1976.

6.24   Kelleher, M.D. and Batts, W.H. Effects of gravity on gas-loaded
       variable conductance heat pipes. ibid.

6.25   Saaski, E.W. and Wilkins, J.S. Heat pipe temperature control util-
       ising a soluble gas absorption reservoir. ibid.

6.26   Marcus, B.D. and Eninger, J.E. Development of vapor-flow-modulation
       variable conductance heat pipes. ibid.

6.27   Roberts, C.C. A variable conductance heat pipe using bubble pump
       injection. ibid.

6.28   Edelstein, F. and Haslett, R. Large variable conductance heat pipe.
       Transverse header. NASA CR-120640, 1975.

Applications of the Heat Pipe

The heat pipe has been, and currently is being, studied for a wide variety
of applications, covering almost the complete spectrum of temperatures
encountered in heat transfer processes. These applications range from the
use of liquid helium heat pipes to aid target cooling in particle
accelerators, to potential developments aimed at new measuring techniques
for the temperature range 2000 - 3000°C.

7.1 Broad Areas of Application

In general the applications come within a number of broad groups, each of
which describes a property of the heat pipe. These groups are:

(i)   Separation of heat source and sink

(ii)   Temperature flattening

(iii)   Heat flux transformation

(iv)   Temperature control

(v)   Thermal diodes and switches

The high effective thermal conductivity of a heat pipe enables heat to be
transferred at high efficiency over considerable distances. In many
applications where component cooling is required, it may be inconvenient or
undesirable thermally to dissipate the heat via a heat sink or radiator
located immediately adjacent to the component. For example, heat dissipation
from a high power device within a module containing other temperature-
sensitive components would be effected by using the heat pipe to connect the
component to a remote heat sink located outside the module. Thermal
insulation could minimise heat losses from intermediate sections of the heat
pipe.

The second property listed above, temperature flattening, is closely related
to source-sink separation. As a heat pipe, by its nature, tends towards
operation at a uniform temperature, it may be used to reduce thermal
gradients between unevenly heated areas of a body. The body may be the outer

255

skin of a satellite, part of which is facing the sun, the cooler sections
being in shadow; this is illustrated diagrammatically in Fig. 7.1.
Alternatively, an array of electronic components mounted on a single pipe
would tend to be subjected to feedback from the heat pipe, creating
temperature equalisation.

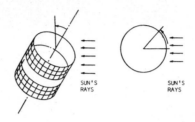

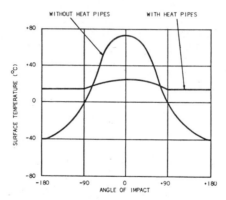

Fig. 7.1   Satellite isothermalisation
(Courtesy Dornier Review)

Heat flux transformation has attractions in reactor technology. In
thermionics, for example, the transformation of a comparatively low heat flux,
as generated by radioactive isotopes, into sufficiently high heat fluxes
capable of being utilised effectively in thermionic generators has been
attempted (7.1).

The fourth area of application, temperature control, is best carried out
using the variable conductance heat pipe. This is fully described in
Chapter 6 and can be used to control accurately the temperature of devices
mounted on the heat pipe evaporator section. To date it has been applied

mainly in spacecraft.

The thermal diode, described in Chapter 5, has a number of specialised applications where heat transport in one direction only is a prerequisite.

As with any other device, the heat pipe must fulfil a number of criteria before it becomes fully acceptable in applications in industry. For example, in
the diecasting and injection moulding industries discussed in the next section,
the heat pipe has to be:

  (i)   Reliable and safe
  (ii)   Satisfy a required performance
  (iii)   Cost-effective
  (iv)   Easy to install and remove

Obviously, each application must be studied in its own right, and the criteria
vary considerably. A feature of the moulding processes, for example, is the
presence of high frequency accelerations and decelerations. In these processes,
therefore, heat pipes should be capable of operating when subjected to this
motion, and this necessitates development work in close association with the
potential users.

7.2 Die Casting and Injection Moulding

Diecasting and injection moulding processes, in which metal alloys or plastics
are introduced in molten form into a die or mould and rapidly cooled to produce a component, often of considerable size and complexity, have enabled mass
production on a considerable scale to be undertaken. The production rate of
very small plastic components may be measured in cycles per second, while alloy castings such as covers for car gearboxes may be produced at upwards of one
per minute. Aluminium, zinc and brass are the most common metals used in die-
cast components, but stainless steel components may now be made using this
technique.

The removal of heat during the solidification process is the most obvious requirement, and nearly all dies are water cooled. However, difficulties are
sometimes experienced in taking water cooling channels to inaccessible parts
of the die. A common solution is to use die inserts made of a more highly conducting material, such as molybdenum, which conducts the heat away to more re-

mote water cooling channels. Furthermore, it is often inconvenient to take
water cooling to movable or removable nozzles, sprue pins, (which fit into the
nozzle where the molten metal is introduced), and cores.

Possibly a more important aspect of die cooling is the need to minimise thermal shock, thus ensuring a reasonable life for the components. With quite large
temperature differences between the molten material and the cooling water,
which must be tolerated by the intervening die, the life of the die can be
shortened. What these parts clearly require is a means of rapidly abstracting
heat from their working surfaces at a temperature more nearly approaching that
of the molten metal.

Two more thermal problems may be mentioned. In some processes it may be necessary or desirable to heat parts of the die to ensure continuous flow of the
molten material to the more inaccessible regions remote to the injection point.
To obtain the subsequent rapid solidification, a change from heating to cooling
is required in a minimum amount of time to keep cycle times as short as possible.

7.2.1 How the heat pipe can assist.

The heat pipe in its simple tubular form
has properties which make it attractive in two areas of application in dies
and moulds. Firstly, the heat pipe may be used to even out temperature gradients in the die by inserting it into the main body of the die, without connecting it to the water cooling circuits. For example, a die used to produce a
cylindrical shaped component may have a significant temperature gradient along
its length. The insertion of heat pipes inside the annular part of the die
could minimise these temperature gradients, as indicated in Fig. 7.2.

Probably the most important application is in assisting heat transfer between
the die face and the water cooling path in areas where hot spots occur. A considerable number of heat pipes manufactured at IRD have been used by Metal
Castings Doehler Limited for this purpose, and significant improvements in production rate have resulted.

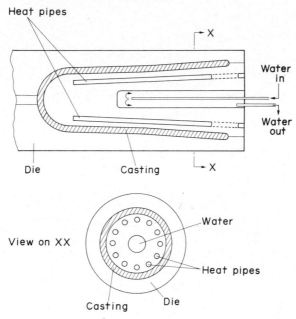

Fig 7.2 The use of heat pipes to reduce die wall
temperature gradients.

In one die a heat pipe was inserted between existing water cooling ducts and
a point closer to the die face, as shown in Fig. 7.3. A substantial improvement in heat removal rate was achieved without increasing the water throughput,
and it was possible to raise the production rate from 25 to 35 shots per hour,
the casting being aluminium and weighing 6 kg. As a bonus, the better thermal
gradient in the die also resulted in improved quality in other parts of the
casting.

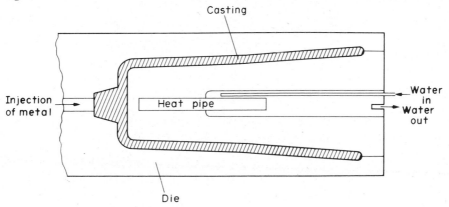

Fig 7.3 The heat pipe used to supplement water
cooling in a die.

Another company who reported the use of heat pipes in aluminium pressure diecasting as long ago as 1972 is Fiat S.p.A. of Turin (7.24).

Fiat carried out successful experiments using heat pipes for rapid cooling of
cores and hot spots, and examples of their use for core cooling are given in
Fig. 7.4, the heat pipe condenser sections being sited in purpose-built water
channels.

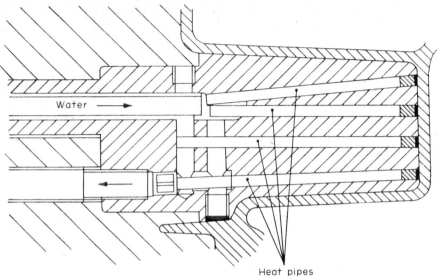

Fig. 7.4. Heat pipes located in cores at Fiat's
diecasting plant.

Other work reported in the literature includes data on the experiments at Bat-
telle's Columbus Laboratory on the use of a heat pipe to transfer heat from the
metal bath to the nozzle on a hot-chamber zinc diecasting machine, (where app-
arently only limited success was achieved). Also the suggestion has been made
that heat pipes, because of their ability to transfer heat in either direction,
could be used to heat dies at start-up, as well as locally cool them (7.25).

Heat pipe core pins have been used in plastics injection moulding, where copper-
beryllium pins, although having a relatively high thermal conductivity when compared with the steel used for the mould, have not removed heat sufficiently
rapidly. In the manufacture of glass containers, work at Stuttgart University
on the use of liquid metal heat pipes to effect cooling in the mould has shown
considerable promise (7.26, 7.27).

In casting and moulding, the range of heat pipe materials and working fluids used is by necessity very wide. Plastics and zinc casting can accept water as the working fluid, as can some aluminium dies. Brass in most cases necessitates a high temperature organic working fluid, while higher melting temperatures lead one into the regime of liquid metal heat pipes (7.28).

7.3 Cooling of Electronic Components

At present possibly the largest application of heat pipes in terms of quality used is the cooling of electronic components such as transistors, other semi-conductor devices, and integrated circuit packages (icp's).

It is convenient to describe in broad terms how the heat pipe can be applied in this area before giving some specific examples. As discussed in Section 7.1, the heat pipe has four main properties, and those of direct interest here are separation of source and sink, temperature flattening, and, more important at present in space electronics cooling, (see Section 7.4), temperature control.

The system geometry may now be considered; this may for convenience be divided into three major categories, each representing a different type of heat pipe system:

(i)   Tubular
(ii)   Flat plate
(iii) Direct contact.

7.3.1 Tubular heat pipes. In its tubular form (with round, oval, rectangular or other cross-sections), two prime functions of the heat pipe may be identi-fied:

(i)   Heat transfer to a remote location
(ii)   Production of a compact heat sink.

By using the heat pipe as a heat transfer medium between two isolated loca-tions, recognisable applications become evident. It becomes possible to con-nect the heat pipe condenser to any of the following:

(i)   A solid heat sink (Fig. 7.5)

(ii)   Immersion in another cooling medium

(iii)   A separate part of the component or component array

(iv)   Another heat pipe

(v)   The wall of the module containing the component(s) being cooled

In applications where size and weight are needed to be kept to a minimum, the near isothermal operation of the heat pipe may be used to raise the temperature of fins or other forms of extended surface. This leads to higher heat transfer to the ultimate sink medium, commonly air, and the advantage may be used to uprate the device or reduce the weight and size of the metal heat sink. There are two possible ways of using the heat pipe here:

(i)   Mount the component directly onto the heat pipe

(ii)   Mount the component onto a plate into which heat pipes are
        inserted.

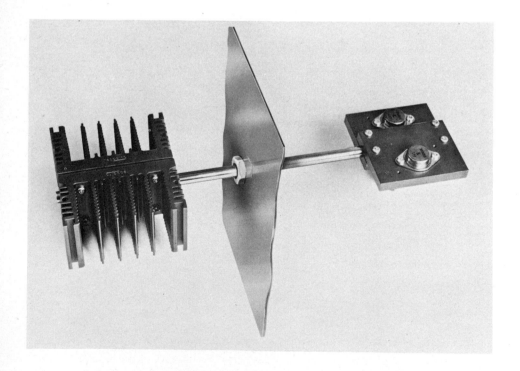

Fig. 7.5   Use of tubular heat pipe to enable heat sink to be some distance from the device being cooled. (Courtesy Redpoint Associates Ltd)

7.3.2 **Flat plate heat pipes.** The second of the three main categories of heat pipes likely to be most useful in electronics cooling is the flat plate unit. It is not envisaged that this will be used in the immediate future to cool very high power units, but its use for temperature flattening and cooling in association with the smaller semiconductor and transistor packages is not in doubt, nor is its application to integrated circuit packages. The applications of the flat plate unit may be summarized thus:

  (i)   Multi-component array temperature flattening (Fig. 7.6)

  (ii)   Multi-component array cooling

  (iii) Doubling as a module wall or mounting plate

Fig. 7.6 Flat plate units for high power dissipation and for providing isothermal surfaces. (Courtesy Solek Ltd)

7.3.3 **Direct contact systems.** One of the problems with integrating heat pipes and the electronic components is that of mounting the devices and minimising interface resistances. Two ways for easing this problem have been proposed and patented in the U.K., (see Appendix 7).

The first of these, developed at Marconi, and shown in Fig. 7.7, involves using a comfortable heat pipe, or more accurately, plate, which can be pressed into intimate contact with heat generating components, with a minimal thermal interface between these and the wick. The second method, proposed by IRD and illustrated in Fig. 7.8, involves removal of the heat pipe wall altogether and the use of a liquid reservoir to feed wicks which cover the components to be cooled. In this case the module would be a sealed unit with provision for heat extraction located externally.

Fig. 7.7   Comformable heat pipe (Courtesy Marconi Co. Ltd)

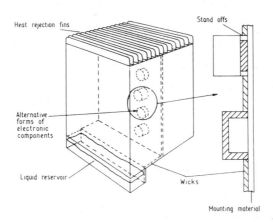

Fig. 7.8   Cooling electronic components by direct contact
(Courtesy IRD)

7.3.4 **Specific applications.** The range of heat pipes used in electronics applications is considerable, some being 1 mm thick by 9 mm wide (7.3), used

for cooling icp's, and others having cross-sections of 25 mm by 12 mm. The
former could transfer 12.5 W, the latter 150 W.

An example of cooling icp's using flat plate 'plug in' heat pipes is that developed by General Electric Company (USA) (7.2). In this type of packaging
system, the thermally critical components were the chip bonding technique and
the air-cooled heat exchanger surface. A number of experiments were carried
out using various chip bonding techniques and an analysis of the overall thermal system performance showed that air or water-cooled systems could cope with
4.4 W/cm^2 input heat flux. For the air-cooled system the junction-to-ambient
thermal resistance was 0.48°C/W, and using water this was reduced to 0.39°C/W.
A similar system using tubular heat pipes is shown in Fig. 7.9.

Fig. 7.9 Cooling of icp's using tubular heat pipes and a
conventional air-cooled heat sink (Courtesy Redpoint Associates Ltd)

Where components may be at a high voltage potential, it may be necessary to
use an electrically insulating heat pipe, and for this the wall, wick and working fluid must be non-conducting. Fig. 7.10 illustrates one form such a unit
might take, and the materials used could be glass, ceramic fibre wicks, and a
fluid such as Flutec, produced by the Imperial Smelting Co. Ltd.

As well as cooling discrete components, heat pipes may be used to remove heat
from only a small part of a device, as in klystron and travelling wavetube
collectors, etc. In addition to cooling applications on the ground and in
space, heat pipes have been used in avionics. Here the major design requirement
stems from the accelerations undergone by the system as the aircraft manoeuvres,
and this can limit performance considerably.

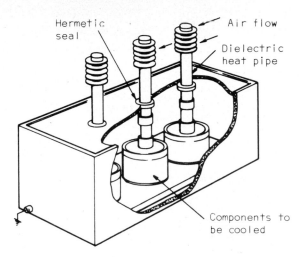

Fig. 7.10 Cooling of high voltage components using electrically insulated heat pipes. (Courtesy Hughes Aircraft Co.)

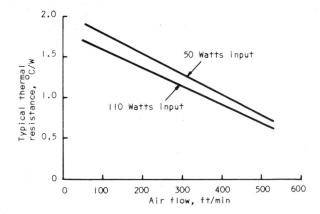

Fig. 7.11   Thermal resistance of finned heat pipes
(Courtesy Solek Ltd.)

An increasing number of companies market or manufacture heat pipes (see Appendix 5) largely directed at electronics cooling, and often are produced with component mounting blocks and finned condensers. Typical performance of units is given in Fig. 7.11.

7.4 Spacecraft

Heat pipes, certainly at vapour temperatures up to 200°C, have probably gained more from developments associated with spacecraft applications than from any other area. The variable conductance heat pipe is a prime example of this 'technological fall-out'. Several types of application may be discussed, and it is convenient to sub-divide these.

7.4.1 Spacecraft temperature equalisation.
Temperature equalisation in spacecraft, whereby thermal gradients in the structure can be minimised to reduce effects of external heating, such as solar radiation, and internal heat generation by electronics components or nuclear power supplies, has been discussed by Savage (7.4) in a paper reviewing several potential applications of the basic heat pipe. The use of a heat pipe connecting two vapour chambers on opposite faces of a satellite is analysed with proposals for reducing the temperature difference between a solar cell array facing the sun, and the cold satellite face. If the solar array was two sided, Savage proposed to mount the cells on a vapour chamber, using one side for radiation cooling, in addition to connecting the cell vapour chamber to extra radiators via heat pipes. Katzoff (7.5) proposed as an alternative the conversion of the tubular structural members of a satellite body into heat pipes.

The use of heat pipes to improve the temperature uniformity of non uniformly irradiated skin structures has been proposed by Thurman and Mei (7.6), who also discuss production of near-isothermal radiator structures utilising heat pipes to improve the efficiency of waste heat rejection, the heat originating in a reactor/thermionic converter system. Conway and Kelly (7.7) studied the feasibility of a continuous circular heat pipe having many combinations of evaporator and condenser surfaces. This was constructed in the form of a toroid with 8 heat sources and 8 heat sinks. They concluded that a continuous heat pipe properly integrated with a spacecraft can be a highly effective means for reducing temperature differentials.

Kirkpatrick and Marcus (7.8) propose the use of heat pipes to implement structural isothermalisation on the National Space Observatory and the Space Shuttle.
It is particularly important to be able to eliminate structural distortions in
orbiting astronomy experiments.

7.4.2 Component cooling, temperature control and radiator design. The widest
application of variable conductance heat pipes and a major use of basic
heat pipe units is in the removal of heat from electronic components and other
heat generating devices on satellites.

The variable conductance heat pipe offers temperature control within narrow
limits, in addition to the simple heat transport function performed by basic
heat pipes. One special requirement cited by Savage (7.4) would arise when it
was required to maintain a particular subsystem at a temperature lower than
that of its immediate environment.

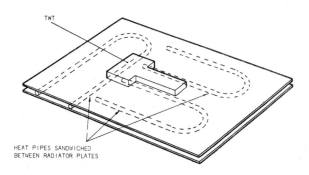

Fig. 7.12 Heat pipes for cooling a TWT mounted on a
radiator plate.

Katzoff (7.5) proposes covering a complete instrument with a wicking material,
and then sealing the instrument within the evaporator section of a heat pipe
which transfers the heat to a radiator. He also suggests that the heat pipe
would be most useful in cases where high intensity localised cooling is required
for only a very short period, where instruments may reach peak power loads for
short parts of a much longer time cycle. Travelling wave tubes (TWT's) are subject to varying internal heat distribution profiles, most of the dissipatiom

being required at the collector. Heat pipes are currently being studied, in
conjunction with a radiator plate, to improve radiator efficiency, and hence
dissipation, from these components. ESRO are currently studying possible configurations of heat pipes in this application. One possible arrangement is
shown in Fig. 7.12.

An interesting form of heat pipe was developed by Basiulis (7.9). The
'unidirectional' heat pipe permits heat flow in one direction, but acts as a
thermal insulator in the opposite direction. Multiple wicks are used to
produce a dry evaporator by limiting fluid return in one direction. By
having a greater number of wicks in the preferred active evaporator, some of
which only extend over the evaporator length, any heat input at the other end
of the pipe results in rapid wick dry out because a considerable portion of
the condensate enters those wick sections which cannot feed liquid back to
the unwanted heat input section. A dielectric heat pipe was used to remove
heat from the TWT in this application, taking heat unidirectionally to
external radiators.

One of the most comprehensive and recent studies on using the VCHP concept
for electronics temperature control has been carried out by Kirkpatrick and
Marcus (7.8). They designed and manufactured a variable conductance pipe,
referred to as the Ames Heat Pipe Experiment (AHPE), in which the heat pipe
provided temperature stability for an On-Board Processor (OBP), by maintaining the OBP platform/AHPE interface at $17 \pm 3^{\circ}C$. Power dissipation from the
electronics processor varied between 10 - 30 W.

Grumman (7.10) investigated other potential spacecraft applications of heat
pipes. This included their use to control the liquid temperature in a closed
circuit environmental control system, using water as the coolant. The
maximum load to be dissipated was 3.82 kW. A second application to thermal
control of equipment was in heat dissipation from a 1 kW power converter
module. Heat pipes were brazed to the rectifier mounting plates. This
resulted in a 15% weight saving and a near isothermal interface mounting
plate. A required dissipation rate of 77 W was achieved.

Kirkpatrick and Marcus quote some very interesting figures on the power
densities encountered in current and proposed systems. For example Apollo
power densities are typically 3 W/linear inch. Present state-of-the-art

electronics for the proposed US Space Station and Earth Orbit Shuttle are of
the order of 30W/linear inch, an order of magnitude higher. Typically a
fluid (liquid) cooled cold rail should handle the power densities
appropriate to Apollo, but to meet post-Apollo requirements, heat pipes are
being given very serious consideration, an augmented heat pipe cold rail
being able to cope with 60W/linear inch. In the Space Shuttle, it is
proposed that heat pipes be used to aid heat transfer from the lubricant
fluid used for the auxiliary power unit.

Turner (7.11), discussing applications for the VCHP, stresses the important
advantage which these (and other heat pipes) offer in permitting direct
thermal coupling of internal spacecraft components to radiators. While with
RCA he developed a VCHP capable of managing power inputs varying between 1
and 65 W. This advantage offered by heat pipes is also mentioned by
Edelstein and Hembach (7.12). Many current electronic packages, when
integrated, rely solely on thermal radiation between the heat source (package)
and heat sink (space) using an intermediate radiating skin which need not
have the components to be cooled in contact with it. Because of the fixed
nature of the thermal coupling, wide ranges of equipment heat generation
and/or environment heat loads will result in large temperature variations.
Turner proposed the VCHP for directly linking sources of heat to the
radiator plate, which would, in turn, be made isothermal.

Scollon (7.13), working at General Electric built a full scale thermal model
of a spacecraft and applied heat pipe technology to many of the thermal
problems. He also selected the Earth Viewing Module of a communications and
navigation satellite as one potential area for application of these devices.
The high solar heat loading on the east and west panels of the satellite
necessitated the use of super-insulation to maintain these surfaces below the
specified upper temperature limits. The north and south faces were the
primary areas for heat rejection. By a system of internal and circumferential
heat pipes, thermal control within the specification was achieved, dealing
with 380 W of internal dissipation and 170 W of absorbed solar energy.

7.4.3 Other applications.

Whilst Cotter was probably the first worker to
note the phenomenon of non-condensable gas generation in a heat pipe, Wyatt
(7.14) was quick to consider their application in space nuclear power sources.

Thurman and Mei (7.6) elaborated on these systems, detailing the use of the
heat pipe in thermionic power supplies as follows:

(i)   Moderator cooling

(ii)   Removal of the heat from the reactor at emitter temperature.
        Each fuel rod would consist of a heat pipe with externally
        attached fuel.

(iii)   Elimination of troublesome thermal gradients along the
        emitter and collector.

Thurman and Mei also designed a boil-off reduction system for stored cryogens
having relatively high boiling temperatures (i.e. liquid oxygen). A closed heat
pipe loop was used to absorb and reject excess heat leakage through the cryogenic tank during extended storage in space. This could result in large weight
savings.

Although not directed at applications in a space environment, the proposals of
Roukis et al (7.10) for cooldown of the Space Shuttle after re-entry are significant and throw light on heat soak problems common to many fields. They
suggested using heat pipes to facilitate rapid cooldown of the Shuttle structure, both prior to and following landing to hasten turn around.

It is advantageous to be able to flight-test heat pipes developed for satellites, to prove that units can withstand the launch environment and meet the
specification in a zero gravity environment. The most recent experiment was
carried out using a heat pipe produced by IKE in Stuttgart. In what became the
'International Heat Pipe Experiment', this unit was launched in late 1974 on
an American rocket, and some zero-gravity operation was monitored before the
rocket returned to earth. The unit was successful, and is the first European
heat pipe to be launched.

A more ambitious programme was conducted in 1972/73 by NASA in conjunction
with Grumman Aerospace Corporation (7.15). This involved collecting flight
data from three 12 mm diameter heat pipes in the form of hoops of 1.22 m diameter mounted in a satellite. The satellite used was the orbiting astronomical
observatory (OAO-C) launched in August 1972. The function of the heat pipes
was to isothermalise the structure onto which the star trackers were mounted;

(these are very sensitive to any structural thermal distortions).

Analysis showed that the circumferential temperature difference in the structure would be reduced by over 75% to about 4°C.

The heat pipes constructed had aluminium walls and ammonia as the working fluid.
Three different wick forms were used. One pipe used simple longitudinal grooves
in the wall, one had a pedestal artery (7.16), and the third heat pipe utilised a spiral artery (7.17). These are fully described in the references cited.
Operating temperatures varied between -20°C and +10°C. Maximum power was up to
90 W. The location of the heat pipe is shown in Fig. 7.13.

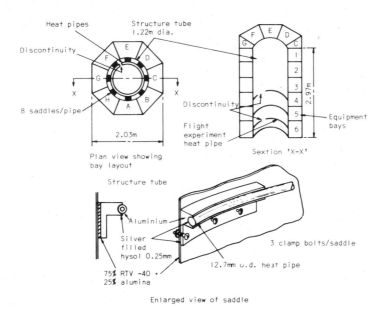

Fig. 7.13 Location of heat pipes on the OAO-C spacecraft,
1973.

The experiment ran for 9 months before the first results were published, and
these showed that flight data obtained from the heat pipes was in excellent
agreement with ground test data. Also the pipes fulfilled their isothermalisation duty and no performance degradation over this period was noted,

Further examples of space applications are given in the bibliography, (Appendix 6.)

7.5 Energy Conservation

The heat pipe, because of its high effectiveness in heat transfer, is a prime
candidate for applications involving the conservation of energy, and has been
used to advantage in heat recovery systems, and energy conversion devices (discussed later in this Chapter).

Energy conservation is becoming increasingly important as the cost of fuel
rises and the reserves diminish, and the heat pipe is proving a particularly
effective tool in a large number of applications associated with conservation.

7.5.1 Heat recovery units. There are a large number of techniques for recover-
ing heat from exhaust air or gas streams, and commercially available systems
fall into two categories - recuperators and regenerators (7.29). The recuperator,
of which the heat pipe heat exchanger is one particular type, functions in
such a way that heat flows steadily and continuously from one fluid to another
through a containing wall.

A heat pipe exchanger consists of a bundle of externally finned tubes which
are made up as individual heat pipes of the type described above. The heat
pipe evaporation/condensation cycle effects the transfer of heat from the "evapor-
ators", located in the duct carrying the counter-current gas stream from which
heat is required to be recovered, to the "condensers" in the adjacent duct
carrying the air which is to be preheated, as illustrated in Fig. 7.14.

Fig. 7.15 shows a unit of this type under test at IRD.

In the heat pipe heat exchanger, the tube bundle may be horizontal, or vertical with the evaporator sections below the condensers. The angle of the heat
pipes may be adjusted "in situ" as a means of controlling the heat transport.
This is a useful feature in air conditioning applications.

Features of heat pipe heat exchangers which are attractive in industrial heat
recovery applications are:

    (i)   No moving parts and no external power requirements, implying high
           reliability

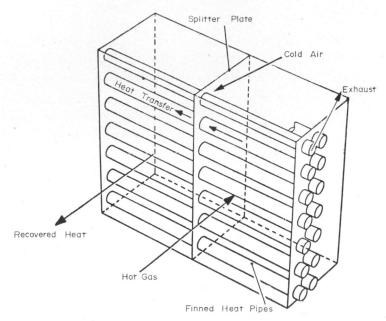

Fig. 7.14 Layout of a heat pipe heat exchanger showing
means of heat transfer.

Fig. 7.15 A heat pipe heat exchanger under test at
International Research & Development Co. Ltd.

(ii)   Cross-contamination is totally eliminated because of a
        solid wall between the hot and cold gas streams

(iii)   A wide variety of sizes are available, and the unit is in general
        compact and suitable for all except the highest temperature appli-
        cations

(iv)   The heat pipe heat exchanger is fully reversible—i.e. heat can be
        transferred in either direction.

(v)   Collection of condensate in the exhaust gases can be arranged, and
        the flexibility accruing to the use of a number of different fin
        spacings can permit easy cleaning if required.

The applications of these devices fall into three main categories:

(i)   Heat recovery in air-conditioning systems, normally involving com-
        paratively low temperatures and duties

(ii)   Recovery of heat from a process exhaust stream to preheat air for
        space heating

(iii)   Recovery of waste heat from processes for reuse in the process, e.g.
        preheating of combustion air. This area of application is the most
        diverse and can involve a wide range of temperatures and duties.

The materials and working fluids used in the heat pipe heat recovery unit depend to a large extent on the operating temperature range and, as far as external tube surface and fins are concerned, on the contamination in the environment in which the unit is to operate. The working fluids for air conditioning and other applications where operating temperatures are unlikely to exceed
$40^{\circ}C$ include the Freons and acetone. Moving up the temperature range, water is
the best fluid to use. For hot exhausts in furnaces and direct gas-fired air
circuits, higher temperature organics can be used.

In most instances the tube material is copper or aluminium, with the same material being used for the extended surfaces. Where contamination in the gas is
likely to be acidic, or at higher temperature where a more durable material
may be required, stainless steel is generally selected.

The tube bundle may be made up using commercially available helically wound
finned tubes, or may be constructed like a refrigeration coil, the tubes being

expanded into plates forming a complete rectangular "fin" running the depth of
the heat exchanger. The latter technique is preferable from a cost point of
view.

Unit size varies with the air flow, a velocity of about 2-4 m/s being generally
acceptable to keep the pressure drop through the bundle to a reasonable level.

Small units having a face size of 0.3 m (height) x 0.6 m (length) are available.
The largest single units are about 5 m in length, and 1.5 m high. The number
of rows of tubes in the direction of the gas flow rarely exceeds eight, and is
most commonly between four and six, although two or three assemblies may be
used in series.

The first heat pipe units developed for this type of application were for heat
recovery, and were made by Q-Dot Corporation of America. Now several other
manufacturers are making similar systems. The heat recovery unit, illustrated
in Fig. 7.16, consists of a bundle of finned heat pipes, having their evaporator sections located in the duct from which heat is to be extracted, and the
condensers in an adjacent duct which may serve as a pre-heating unit. Such a
system is claimed to be up to 70% efficient, and the savings in energy may enable the system to pay for itself in a period of as little as two years.

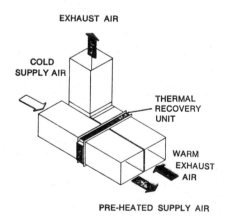

Fig. 7.16 Air-to-air heat pipe recovery unit used for pre-
heating air. (Courtesy Q-Dot Corporation)

As well as recovering heat, the heat pipes can be used to pre-cool supply air
by rejecting heat to a cold exhaust duct, as for example in an air-conditioning

application. A typical unit may have a face area per duct of 4 m x 2 m, and would contain up to several hundred heat pipes.

To demonstrate the effect of one of these units the following example is taken from the Isothermics brochure (see Appendix 5).

Application   An oil fired drying oven exhausts $64m^3$/min air at 200°C. Heat input is 500 kW. It is required to heat make-up air to 20°C in winter and to heat process air to 95°C in summer. Oven operating 21 hrs per day. Allowable pressure drop 2.2 mm Hg.

System Design   Heat pipe heat exchanger with 1.0 x 1.3 m face area, 4 tube rows deep. Tubes 19mm o.d. with fin height 9.5 mm. All copper with nickel plate protective coating. Working fluids methanol and water. Weight 300 kg; total installation cost £10 000.

Results   Effective recovery   56%
           Annual fuel saving $8000.

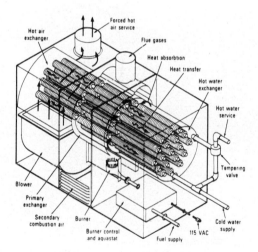

Fig. 7.17 Heat pipe heat exchanger providing hot water and ducted warm air for domestic use. (Courtesy Isothermics Inc.)

The heat pipe is not restricted to heat recovery applications. Its use in conjunction with direct heating such as a gas burner is under investigation, and

a system which can provide directed warm air for space heating and hot water
for other services is shown in Fig. 7.17. The heat pipes used in this application have the evaporator in the centre and a condenser section at each end.
A gas burner provides heat, being thermostatically controlled in such a way
as to maintain the hot water temperature at 55 - 80°C. An air blower is situated below the other condenser section to provide distribution of warm air for
space heating.

It is claimed that the unit can heat 15 - 20 gallons of water per hour. In
winter the hot water can serve as a thermal storage medium when the gas burner
is not in operation, this section becoming the evaporator to supply heat to the
air duct. In summer an adsorption type air conditioning unit could be added
which would utilise the cold water as a heat sink.

The manufacturers estimate that the basic unit will sell for less than $1000.

Fig. 7.18 A compact heat pipe heat exchanger for heat
           transfer between air and water. (Courtesy
           Solek Ltd.)

Also available is the 'Air-O-Space' heater, which can be added to any domestic
boiler flue, extracting heat to provide warm air distribution via a built-in
fan. The unit, it is claimed, can recover in excess of 4 kW, depending on the
exhaust flue gas temperature.

A small heat pipe heat exchanger of the water-to-air type, manufactured in the
United Kingdom, is illustrated in Fig. 7.18

7.6 Thermionic Power Generation

The thermionic generation method has received a considerable amount of attention as a possible way of converting nuclear fission heat directly to electricity. It has particular advantages for space applications at high power levels
of a few megawatts and above. The thermionic generator can be situated either
inside or outside the nuclear reactor core. There are nuclear and other advantages in the latter arrangement but the extraction of the nuclear generated
heat at a temperature of around $1600^{\circ}C$ presents a serious problem which it has
been suggested might be solved by the use of heat pipes. The basic generator
is shown in Fig. 7.19. Electrons are emitted from the emitter and cross the
interelectrode space to be collected at the cooler collector surface; the electrons return to the emitter through the external electrical load resistance.
The device is a heat engine which converts some of the thermal energy supplied
to the emitter into electrical energy and rejects the remainder at a lower temperature at the collector.

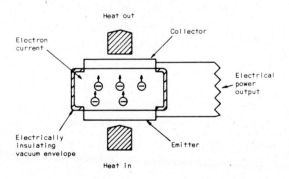

Fig. 7.19 The thermionic generator

HP—J

In the simple generator shown, it is necessary for the electrodes to be spaced
very close together, typically 0.005 mm in order to limit the free electron
space charge which tends to limit the current which may be drawn from the generator. It is usual to arrange for a low pressure caesium vapour to fill the
electrode spacing, the caesium readily forms positive ions which neutralise
the space charge and enable a spacing of around 0.5 mm to be used. Thermionic
generators are made with either cylindrical, coaxial or parallel plane electrodes.

Typical characteristics of a thermionic generator are given in Table 7.1.

TABLE 7.1   CHARACTERISTICS OF THERMIONIC GENERATORS

Emitter temperature ($^\circ$C)	1600
Collector temperature ($^\circ$C)	600
Output voltage (V)	0.3 - 0.8
Output current density (A/cm^2)	3 - 20
Output power density (W/cm^2)	3 - 15
Efficiency (%)	10 - 20

The heat flux at the electrodes is high, in the range of 30 W/cm^2 to 150 W/cm^2.
Generators of this type have no moving parts and operate reliably for long periods. The high reject temperature is particularly suitable for use with radiation cooling at high temperatures.

Heat pipes can be used both for connecting the thermionic generator to the
heat source and also for connecting the generator to the heat rejecting radiator. Because of the importance of reliability and lifetime in space applications a great deal of work on fabrication techniques, materials compatability
and life testing has been carried out on suitable high temperature heat pipes
both in the United States and in Europe. Lithium and silver have been considered for conducting heat to the emitter and sodium and potassium for cooling
the collector. (Compatibility data is presented in Chapter 3).

7.7 Preservation of Permafrost

Probably the largest contract for heat pipes to date was placed with McDonnell
Douglas Corporation by Alyeska Pipeline Service Company for nearly 100,000 heat

pipes for the Trans-Alaska pipeline. The value of the contract is approximately
$13 000 000.

The function of these units is to prevent thawing of the permafrost around the
pipe supports for elevated sections of the pipeline. Diameters of the heat
pipes used are 5 and 7.5 cm, and lengths vary between 9 and 18 m.

Several constructional problems are created when foundations have to be on or
in a permafrost region. Frost heaving can cause differential upward motion of
piles resulting in severe structural distortions. On the other hand, if the
foundations are in soil overlying permafrost, downward movements can occur.

If the thermal equilibrium is disturbed by operation of the pipeline, thawing
of the permafrost could occur, with the active layer (depth of annual thaw) increasing each summer until a new thermal equilibrium is achieved. This will of
course affect the strength of the ground and the integrity of any local found-
ation structure. Insulation, ventilation and refrigeration systems have already
been used to maintain permafrost around foundations. The use of heat pipes and
thermosyphons has also been studied by several laboratories (7.20, 7.21).

The system developed by McDonnell Douglas (7.20) uses ammonia as the working
fluid, heat from the ground being transmitted upwards to a radiator located
above ground level. Several forms of 'Cryo-anchor' have been tested, and two

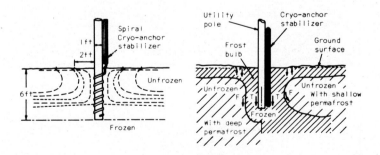

Fig. 7.20 Types of 'Cryo-anchor' for permafrost preservation. (Courtesy McDonnell Douglas Corp.)

types are illustrated in Fig. 7.20. As the heat pipe is very long and is operating in the vertical position, the only means for transferring heat from the
atmosphere into the ground will be conduction along the solid wall.

Results have shown that rapid soil cooling occurs in the autumn after installation and as air temperatures increased in the spring, Cryo-anchor cooling
ceased. The temperature rose fairly rapidly as heat flowed from the surrounding less cold permafrost, with large radial temperature gradients. As the
gradients became smaller, the temperature rise slowed through the summer months.

At the end of the thaw season, the permafrost remained almost $0.5^{\circ}C$ below its
normal temperature. This cooling would permit a 10 - 12 m reduction in pile
length required to support the pipe structure. The system is now installed on
the pipeline.

7.8 Stirling Engines

The Stirling engine is a reciprocating external combustion engine employing
gas as the working fluid. Like all heat engines, it has a high-temperature and
a low-temperature heat exchanger. A heat pipe may be used to transport heat
from a single heat source to the separate cylinders of a multi-cylinder engine.
Heat pipes could also be used to transport rejected heat to the radiator. The
ideal Stirling cycle is shown in Fig. 7.21.

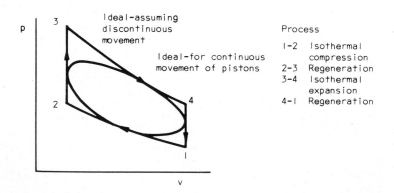

Fig. 7.21 Stirling cycle (ideal)

Referring to the pressure volume diagram (Fig. 7.21):

Gas is compressed at constant temperature from state 1 to state 2. Heat is
given out during this process. The gas is then heated from state 2 to a higher
temperature at state 3. The gas is now allowed to expand at constant temperature to state 4; during this process heat will be adsorbed and useful work done.
The gas is now cooled at constant volume from state 4 to state 1 to complete
the cycle.

For an ideal engine the net work output will be the difference between the expansion work and the compression work. The heat given out in the constant volume cooling process is stored in a regenerator and used to reheat the gas on
the second constant volume process. Since heat is only abstracted from the high
temperature source, and given up to the low temperature sink in the two isothermal processes, the cycle has the maximum theoretical efficiency for a heat
engine.

The weight, size and efficiency of a Stirling engine are similar to those of a
diesel engine of the same power and speed. The Stirling engine has, however,
certain advantages over the diesel, including those of long unattended life,
low exhaust emissions and low noise and vibration.

In order to extract maximum power from a given engine, a working fluid having
a high heat transfer capability is required, normally hydrogen or helium, and
must be operated at high temperature and high pressure levels, typical values
being $700 - 750^{\circ}C$, and 100 atm. The sodium/stainless steel heat pipe is particularly suited for the high temperature end of the Stirling engine and such
systems have been developed at both Philips and Reading University. The stainless steel container is suitable for flame heating. Fig. 7.22 shows a typical
arrangement which in this case makes use of a fluidised bed as the heat source.

7.9 The Vapipe

The National Engineering Laboratory and the Shell Research Laboratory have co-
operated in applying the heat pipe to the problem of exhaust emissions from
petrol engines. Fig. 7.23 shows how the carbon monoxide CO, unburnt hydrocar-
bons $H_x C_y$ and oxides of nitrogen NO_x content of the exhaust will vary with air
to fuel ratio in a conventional car engine. Maximum efficiency is achieved at

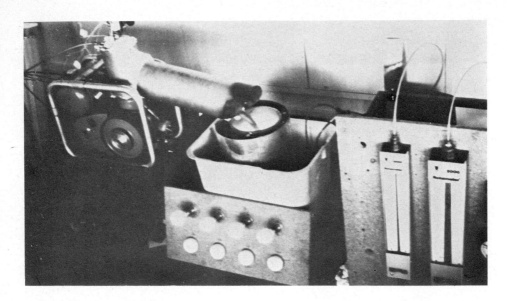

Fig. 7.22   Stirling engine heated using fluidised bed and connecting heat pipe.

around 15:1 and maximum power at 12:1. It is seen from Fig. 7.23 that as the air fuel ratio is increased the CO content decreases and the H_xC_y and NO_x go through a minimum and a maximum respectively. Considerable improvement in both CO and NO_x content could be achieved by selecting a very weak mixture, but this is not possible in a standard engine/carburettor system due to ignition difficulty. Ignition problems arise because the fuel is not fully vaporised; this has two effects, the fuel is not distributed equally between the cylinders and the vapour content is not as high as it should be because of the presence of liquid fuel. In the Vapipe or Heat Pipe Vaporiser Unit a heat pipe is used to couple the exhaust thermally to the induction manifold at the carburettor outlet. By this means heat from the exhaust is transferred to the fuel air mixture causing full evaporation (see Fig. 7.24). It is found that under these conditions a mixture as lean as 22:1 will ignite without difficulty. The resulting improvement in NO_x and CO is indicated in Fig. 7.23.

NEL/Shell report tests with a 1.8 l. engine in which 2.5 kW of heat are required for evaporation when the engine is at full throttle. In one design the whole of the air passes through the vaporising unit; in a later design 20% of the air passes, with the fuel, through the vaporiser and the remaining 80% of the air bypasses the unit.

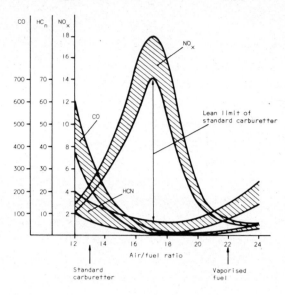

Fig. 7.23   The Vapipe - typical variation of emissions with air/fuel ratio.

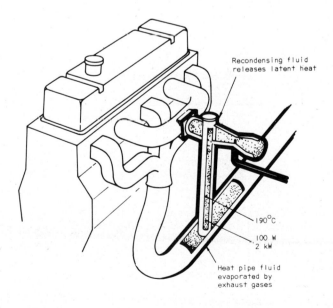

Fig. 7.24   The Vapipe installation.

7.10   A Biological Heat Pipe

Although the phenomena discussed below are only at present the basis of a working hypothesis, and further study is being carried out, it is felt that the
hypothesis is of sufficient interest to be included in this Chapter. The hypothesis put forward (7.22, 7.23) is that there is an analogy between the functioning of a sweat gland and the operation of a heat pipe. Several assumptions
have been made concerning the resting sweat gland, based on observations of their
behaviour. Most important is the assumption that the resting eccrine sweat
gland functions in thermoregulation, and hence water is being continuously se-
creted by the coils. In order to provide efficient cooling this water must be
evaporated near the base of the duct. Other observations detailed in reference
7.22 lead one to deduce that a refluxing of water back down the sweat ducts
exists, assisted by capillary attraction in the mucilaginous lining, and by
osmosis. (The resting state occurs without secretion of liquid onto the skin
surface, which is active sweating). The model presented, namely evaporation
of water, recondensation and flow of the condensate by, in part, capillary action back to the 'evaporator' has an equivalent in engineering heat transfer,
namely the heat pipe.

This analogy is illustrated in Fig. 7.25.

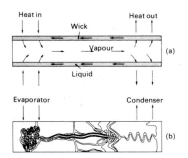

Fig. 7.25   Diagrammatic representation of a heat pipe (a)
and an eccrine sweat gland (b).

Heat pipe theory was used to predict the performance of sweat glands, and the results compared with 'in vivo' measurements of body heat loss. An 'equivalent heat pipe' was defined as follows:

Heat pipe length   -   5 mm
Heat pipe diameter   -   0.018 mm
Wick thickness   -   0.001 mm
Wick pore size   -   0.00006 mm
Wick thermal conductivity   -   0.6 W/m^2 $^{\circ}$C (Value for water)
Porosity of wick   -   69%
Mode of heat transfer into heat pipe - conduction
Length of evaporator   -   2.5 mm
Temperature around evaporator   -   40°C
Vapour temperature   -   25-40°C in 5° increments
External temperature   -   20°C
Working fluid   -   water
Angle of heat pipe to horizontal -   various

In applying this model to the sweat gland the assumption was made that the vapour pressure generated in the duct is sufficient to purge the duct of air. Based on observations made (7.22, 7.23) on sweat duct behaviour, this seems reasonable. (In some cases, i.e. skin grafts, sweat ducts are completely sealed, but still transport significant quantities of heat.)

The mathematical model is a simplification in that liquid transport to the evaporator is assumed to occur solely through the wick. Liquid flowing into the base of the sweat duct from other surrounding regions could affect the wicking limit, raising the capability, but if this flow is also a capillary action through pores of similar orders of magnitude to those in the mucilaginous lining, the results calculated by the program should be representative to a first order approximation.

The results showed that the wick capillary pumping capability limited the performance of the 'equivalent heat pipe' to an axial heat transport capability of 1×10^{-5}W.

The skin density of sweat glands is of the order of 200/cm^2, and the surface area of the body is approximately 2 m^2.

Integrating over the whole body, assuming that all sweat glands have the same
size and temperature and therefore the same performance, a value may be obtained for the total heat dissipation by the resting sweat glands. Using the
above sweat gland density and body surface area, this figure is 34 W.   Total
heat loss from the body sedentary position has independently been measured to be
approximately 100 W.

Further work involving the investigation of electro-osmotic effects and diffusion in small capillaries is being carried out.

7.11 Other Applications

It has been possible to describe in detail only a few of the many applications
of heat pipes, and for a more detailed treatment the reader is referred to
Appendix 6. However, the use of heat pipes in solar collectors and as cooking
pins, acting as sophisticated meat skewers (see Fig. 7.26), are but two of the
many applications of these devices. Other range from aircraft brake cooling
to gas turbine recuperation, and from cryoprobes to furnace muffle tubes, some
of which are illustrated in Figs. 7.27 - 7.29 (The devices illustrated in these
figures have been developed by Hughes Aircraft Company).

Fig 7.26 A heat pipe cooking pin - used to
          speed up roasting of meat, etc.

Fig. 7.27 Heat pipe griddle with liquid fuel burner assembly.
          The griddle can maintain the surface isothermal (to
          within 2°C) at temperatures between 100°C and 232°C.
          Dowtherm A is used in this unit.(Hughes Aircraft Co.)

Fig. 7.28 Kryostik: A cryosurgical instrument utilizing an open
          loop heat pipe and superinsulation. Liquid nitrogen is
          the working fluid.(Hughes Aircraft Co.)

Fig. 7.29 (a)   Exploded view of a microwave tube incorporating an
electrically insulated dielectric heat pipe.

Fig. 7.29 (b)   The assemble microwave tube.   Heat pipe duty is 1.5 kW
and electrical insulation 15 kV, using a Refrasil and
copper felt wick structure.   (Hughes Aircraft Co).

NOTE: The devices shown in Figs. 7.27-7.29 were developed by Hughes Aircraft
Company, Torrance, California.

REFERENCES

7.1   Leefer, B.I. Nuclear thermionic energy converter. Proceedings of 20th
        Annual Power Sources Conf., May 1966, pp 172 - 175.

7.2   Corman, J.C. and McLaughlin, M.H. Thermal development of heat pipe
        cooled I.C. packages. ASME Paper 72-WA/HT-44, 1972.

7.3   Basiulis, A. and Hummel, T.A. The application of heat pipe tech-
        niques to electronic component cooling. ASME Paper 72-WA/HT-42, 1972.

7.4   Savage, C.J. Heat pipes and vapour chambers for satellite thermal
        balance. RAE Tech. Report 69125, June 1969.

7.5   Katzoff, S. Heat pipes and vapour chambers for thermal control of
        spacecraft. AIAA Paper 67-310, April, 1967.

7.6   Thurman, J.L. and Mei, S. Application of heat pipes to spacecraft
        thermal control problems. Tech. Note AST-275, Brown Engineering
        (Teledyne), July, 1968.

7.7   Conway, E.C. and Kelley, M.J. A continuous heat pipe for spacecraft
        thermal control. General Electric Space Systems, Pennsylvania (un-
        dated).

7.8   Kirkpatrick, J.P. and Marcus, B.D. A variable conductance heat pipe
        experiment. AIAA Paper 71-411, 1971.

7.9   Basiulis, A. Uni-directional heat pipes to control TWT temperature
        in synchronous orbit. NASA Contract NAS-3-9710, Hughes Aircraft Co.,
        California, 1969.

7.10   Roukis, J. et al. Heat pipe applications for space vehicles. Grum-
        man Aerospace, AIAA Paper 71-412, 1971.

7.11   Turner, R.C. The constant temperature heat pipe - a unique device
        for the thermal control of spacecraft components. AIAA 4th Thermo-
        physics Conf., Paper 69-632, (RCA), June, 1969.

7.12   Edelstein, F. and Hembach, R.J. Design, fabrication and testing of
        a variable conductance heat pipe for equipment thermal control.
        AIAA Paper 71-422, 1971.

7.13   Scollon, T.R. Heat pipe energy distribution system for spacecraft
        thermal control. AIAA Paper 71-412, 1971.

7.14   Wyatt, T. Controllable heat pipe experiment. Johns Hopkins Univ.,
        Applied Physics Lab. Report SDO-1134, March, 1965.

7.15   Harwell, W. et al. Orbiting astronomical observatory heat pipe
        flight performance data. AIAA Paper 73-758, 1973.

7.16   Bienert, W. and Kroliczek, E. Experimental high performance heat
        pipes for the OAO-C spacecraft. ASME Paper 71-Av-26, July, 1971.

7.17   Edelstein, F. et al. Development of a self-priming high capacity
        heat pipe for flight on OAO-C. AIAA Paper 72-257, 1972.

7.18   Deyoe, D.P. Heat recovery - how can the heat pipe help. ASHRAE Jour-
        nal, pp 35 - 38, April, 1973.

7.19   Behrens, C.W. Heat pipes: Breakthrough in thermal economy? Appliance
        manufacturer (US), pp 72 - 75, Nov. 1973.

7.20   Waters, E.D. Arctic tundra kept frozen by heat pipes. The Oil and
        Gas Journal (US), pp 122 - 125, Aug. 26, 1974.

7.21   Larkin, B.S. and Johnston, G.H. An experimental field study of the
        use of two-phase thermosiphons for the preservation of permafrost.
        Paper at 1973 Annual Congress of Engineering. Inst. of Canada, Mon-
        treal, Oct. 1973.

7.22   Thiele, F.A.J., Mier, P.D. and Reay, D.A. Heat transfer across the
        skin: The role of the resting sweat gland. Proceedings Congress on
        Thermography, Amsterdam, Holland, June, 1974.

7.23   Thiele, F.A.J. Measurements on the surface of the skin. Doctoral
        thesis, Dept. of Dermatology, Nijmegen University, Holland, 1974.

7.24   Mascaretti, F.C. and Medana, R. Diecasting parts for Fiat front
        drive cars. Proc. 7th SDCE Int. Diecasting Congress, Paper No. 4272,
        Chicago, Oct. 1972.

7.25   Winship, J. Diecasting sharpens its edge. American Machinist, Vol.
        118, pp 77-88, Nov. 1974.

7.26   U.K. Patent 1,309,911. Control of the temperature of tools for glass-
        forming machines. Herman Heye K.G., Germany. Filed 2nd March 1970.

7.27   Brost, O. et al. Industrial applications of alkali-metal heat pipes.
        Proc. 1st Int. Heat Pipe Conference, Paper 11-3, Stuttgart, Oct.
        1973.

7.28   Reay, D.A. Heat pipes - a new diecasting aid. Proc. 1st National
        Diecasters Conference, Birmingham Exhibition Centre, May 1977.

7.29   Reay, D.A. Industrial Energy Conservation. Pergamon Press, Oxford,
        1977.

Appendix 1
Working Fluid Properties

(See also Table 3.1)

Fluids listed:	Helium	Heptane
(in order of	Nitrogen	Water
appearance).	Ammonia	Flutec PP9
	Freon 11	Thermex
	Pentane	Mercury
	Freon 113	Caesium
	Acetone	Potassium
	Methanol	Sodium
	Flutec PP2	Lithium
	Ethanol	

Properties listed:
- Latent heat of evaporation
- Liquid density
- Vapour density
- Liquid thermal conductivity
- Liquid dynamic viscosity
- Vapour dynamic viscosity
- Vapour pressure
- Vapour specific heat
- Liquid surface tension

HELIUM

Temp. °C	Latent Heat kJ/kg	Liquid Density kg/m^3	Vapour Density kg/m^3	Liquid Thermal Conductivity W/m°C	Liquid Viscos. cPx10^2	Vapour Viscos. cPx10^3	Vapour Press. Bar	Vapour Specific Heat kJ/kg°C	Liquid Surface Tension N/mx10^3
-271	22.8	148.3	26.0	1.81	3.90	0.20	0.06	2.045	0.26
-270	23.6	140.7	17.0	2.24	3.70	0.30	0.32	2.699	0.19
-269	20.9	128.0	10.0	2.77	2.90	0.60	1.00	4.619	0.09
-268	4.0	113.8	8.5	3.50	1.34	0.90	2.29	6.642	0.01

P.D. DUNN   D.A. REAY

NITROGEN

Temp.	Latent Heat	Liquid Density	Vapour Density	Liquid Thermal Conductivity	Liquid Viscos.	Vapour Viscos.	Vapour Press.	Vapour Specific Heat	Liquid Surface Tension
$°C$	kJ/kg	kg/m^3	kg/m^3	$W/m°C$	$cPx10^1$	$cPx10^2$	Bar	$kJ/kg°C$	$N/mx10^2$
-203	210.0	830.0	1.84	0.150	2.48	0.48	0.48	1.083	1.054
-200	205.5	818.0	3.81	0.146	1.94	0.51	0.74	1.082	0.985
-195	198.0	798.0	7.10	0.139	1.51	0.56	1.62	1.079	0.870
-190	190.5	778.0	10.39	0.132	1.26	0.60	3.31	1.077	0.766
-185	183.0	758.0	13.68	0.125	1.08	0.65	4.99	1.074	0.662
-180	173.7	732.0	22.05	0.117	0.95	0.71	6.69	1.072	0.561
-175	163.2	702.0	33.80	0.110	0.86	0.77	8.37	1.070	0.464
-170	152.7	672.0	45.55	0.103	0.80	0.83	10.07	1.068	0.367
-160	124.2	603.0	80.90	0.089	0.72	1.00	19.37	1.063	0.185
-150	66.8	474.0	194.00	0.075	0.65	1.50	28.80	1.059	0.110

AMMONIA

$°C$	kJ/kg	kg/m^3	kg/m^3	$W/m°C$	cP	$cPx10^2$	Bar	$kJ/kg°C$	$N/mx10^2$
-60	1434	714.4	0.03	0.294	0.36	0.72	0.27	2.050	4.062
-40	1384	690.4	0.05	0.303	0.29	0.79	0.76	2.075	3.574
-20	1338	665.5	1.62	0.304	0.26	0.85	1.93	2.100	3.090
0	1263	638.6	3.48	0.298	0.25	0.92	4.24	2.125	2.480
20	1187	610.3	6.69	0.286	0.22	1.01	8.46	2.150	2.133
40	1101	579.5	12.00	0.272	0.20	1.16	15.34	2.160	1.833
60	1026	545.2	20.49	0.255	0.17	1.27	29.80	2.180	1.367
80	891	505.7	34.13	0.235	0.15	1.40	40.90	2.210	0.767
100	699	455.1	54.92	0.212	0.11	1.60	63.12	2.260	0.500
120	428	374.4	113.16	0.184	0.07	1.89	90.44	2.292	0.150

FREON 11

Temp.	Latent Heat	Liquid Density	Vapour Density	Liquid Thermal Conductivity	Liquid Viscos.	Vapour Viscos.	Vapour Press.	Vapour Specific Heat	Liquid Surface Tension
$^{\circ}C$	kJ/kg	kg/m^3	kg/m^3	W/m$^{\circ}$C	cP	cPx10^2	Bar	kJ/kg$^{\circ}$C	N/mx10^2
-60	211.9	1672	0.04	0.121	1.19	0.86	0.02	0.476	2.95
-40	204.0	1622	0.04	0.115	0.98	0.88	0.05	0.497	2.70
-20	196.8	1578	1.04	0.111	0.70	0.95	0.16	0.516	2.40
0	190.0	1533	2.59	0.108	0.55	1.01	0.42	0.532	2.18
20	183.4	1487	5.38	0.100	0.44	1.08	0.93	0.546	1.92
40	175.6	1439	10.07	0.097	0.37	1.14	1.82	0.561	1.66
60	167.5	1389	16.85	0.094	0.32	1.20	3.14	0.576	1.40
80	159.0	1334	30.56	0.089	0.28	1.25	5.85	0.590	1.14
100	146.9	1265	49.04	0.076	0.25	1.31	9.53	0.607	0.90
120	134.4	1194	67.53	0.064	0.23	1.37	13.21	0.623	0.63
140	117.0	1105	110.66	0.055	0.22	1.49	18.92	0.646	0.37

PENTANE

$^{\circ}C$	kJ/kg	kg/m^3	kg/m^3	W/m$^{\circ}$C	cP	cPx10^2	Bar	kJ/kg$^{\circ}$C	N/mx10^2
-20	390.0	663.0	0.01	0.149	0.344	0.51	0.10	0.825	2.01
0	378.3	644.0	0.75	0.143	0.283	0.53	0.24	0.874	1.79
20	366.9	625.5	2.20	0.138	0.242	0.58	0.76	0.922	1.58
40	355.5	607.0	4.35	0.133	0.200	0.63	1.52	0.971	1.37
60	342.3	585.0	6.51	0.128	0.174	0.69	2.28	1.021	1.17
80	329.1	563.0	10.61	0.127	0.147	0.74	3.89	1.050	0.97
100	295.7	537.6	16.54	0.124	0.128	0.81	7.19	1.088	0.83
120	269.7	509.4	25.20	0.122	0.120	0.90	13.81	1.164	0.68

FREON 113

Temp.	Latent Heat	Liquid Density	Vapour Density	Liquid Thermal Conductivity	Liquid Viscos.	Vapour Viscos.	Vapour Press.	Vapour Specific Heat	Liquid Surface Tension
$^{\circ}$C	kJ/kg	kg/m^3	kg/m^3	W/m$^{\circ}$C	cP	cPx10^2	Bar	kJ/kg$^{\circ}$C	N/mx10^2
-50	173.0	1720	0.15	0.120	2.300	0.85	0.01	0.600	2.86
-30	167.8	1683	0.32	0.119	1.604	0.90	0.03	0.613	2.60
-20	165.4	1664	0.46	0.118	1.323	0.92	0.05	0.619	2.47
-10	163.2	1643	0.77	0.118	1.108	0.94	0.09	0.626	2.34
0	160.6	1621	1.26	0.117	0.942	0.97	0.12	0.632	2.21
10	158.0	1599	1.95	0.108	0.812	0.99	0.19	0.644	2.08
20	155.2	1576	3.00	0.098	0.707	1.02	0.37	0.656	1.96
30	152.3	1553	4.34	0.097	0.622	1.04	0.55	0.664	1.84
40	149.2	1529	6.02	0.095	0.553	1.07	0.79	0.669	1.73
50	145.9	1503	8.79	0.094	0.502	1.09	1.11	0.674	1.62
70	139.4	1452	14.34	0.091	0.401	1.13	2.04	0.691	1.40

ACETONE

$^{\circ}$C	kJ/kg	kg/m^3	kg/m^3	W/m$^{\circ}$C	cP	cPx10^2	Bar	kJ/kg$^{\circ}$C	N/mx10^2
-40	660.0	860.0	0.03	0.200	0.800	0.68	0.01	2.00	3.10
-20	615.6	845.0	0.10	0.189	0.500	0.73	0.03	2.06	2.76
0	564.0	812.0	0.26	0.183	0.395	0.78	0.10	2.11	2.62
20	552.0	790.0	0.64	0.181	0.323	0.82	0.27	2.16	2.37
40	536.0	768.0	1.05	0.175	0.269	0.86	0.60	2.22	2.12
60	517.0	744.0	2.37	0.168	0.226	0.90	1.15	2.28	1.86
80	495.0	719.0	4.30	0.160	0.192	0.95	2.15	2.34	1.62
100	472.0	689.6	6.94	0.148	0.170	0.98	4.43	2.39	1.34
120	426.1	660.3	11.02	0.135	0.148	0.99	6.70	2.45	1.07
140	394.4	631.8	18.61	0.126	0.132	1.03	10.49	2.50	0.81

METHANOL

Temp.	Latent Heat	Liquid Density	Vapour Density	Liquid Thermal Conductivity	Liquid Viscos.	Vapour Viscos.	Vapour Press.	Vapour Specific Heat	Liquid Surface Tension
$^{\circ}C$	kJ/kg	kg/m^3	kg/m^3	W/m$^{\circ}$C	cP	cPx10^2	Bar	kJ/kg$^{\circ}$C	N/mx10^2
-50	1194	843.5	0.01	0.210	1.700	0.72	0.01	1.20	3.26
-30	1187	833.5	0.01	0.208	1.300	0.78	0.02	1.27	2.95
-10	1182	818.7	0.04	0.206	0.945	0.85	0.04	1.34	2.63
10	1175	800.5	0.12	0.204	0.701	0.91	0.10	1.40	2.36
30	1155	782.0	0.31	0.203	0.521	0.98	0.25	1.47	2.18
50	1125	764.1	0.77	0.202	0.399	1.04	0.55	1.54	2.01
70	1085	746.2	1.47	0.201	0.314	1.11	1.31	1.61	1.85
90	1035	724.4	3.01	0.199	0.259	1.19	2.69	1.79	1.66
110	980	703.6	5.64	0.197	0.211	1.26	4.98	1.92	1.46
130	920	685.2	9.81	0.195	0.166	1.31	7.86	1.92	1.25
150	850	653.2	15.90	0.193	0.138	1.38	8.94	1.92	1.04

FLUTEC PP2

$^{\circ}C$	kJ/kg	kg/m^3	kg/m^3	W/m$^{\circ}$Cx10	cP	cPx10	Bar	kJ/kg$^{\circ}$C	N/mx10^2
-30	106.2	1942	0.13	0.637	5.200	0.98	0.01	0.72	1.90
-10	103.1	1886	0.44	0.626	3.500	1.03	0.02	0.81	1.71
10	99.8	1829	1.39	0.613	2.140	1.07	0.09	0.92	1.52
30	96.3	1773	2.96	0.601	1.435	1.12	0.22	1.01	1.32
50	91.8	1716	6.43	0.588	1.005	1.17	0.39	1.07	1.13
70	87.0	1660	11.79	0.575	0.720	1.22	0.62	1.11	0.93
90	82.1	1599	21.29	0.563	0.543	1.26	1.43	1.17	0.73
110	76.5	1558	34.92	0.550	0.429	1.31	2.82	1.25	0.52
130	70.3	1515	57.21	0.537	0.314	1.36	4.83	1.33	0.32
160	59.1	1440	103.63	0.518	0.167	1.43	8.76	1.45	0.01

ETHANOL

Temp.	Latent Heat	Liquid Density	Vapour Density	Liquid Thermal Conductivity	Liquid Viscos.	Vapour Viscos.	Vapour Press.	Vapour Specific Heat	Liquid Surface Tension
$^\circ$C	kJ/kg	kg/m^3	kg/m^3	W/m$^\circ$C	cP	cPx10^2	Bar	kJ/kg$^\circ$C	N/mx10^2
-30	939.4	825.0	0.02	0.177	3.40	0.75	0.01	1.25	2.76
-10	928.7	813.0	0.03	0.173	2.20	0.80	0.02	1.31	2.66
10	904.8	798.0	0.05	0.170	1.50	0.85	0.03	1.37	2.57
30	888.6	781.0	0.38	0.168	1.02	0.91	0.10	1.44	2.44
50	872.3	762.2	0.72	0.166	0.72	0.97	0.29	1.51	2.31
70	858.3	743.1	1.32	0.165	0.51	1.02	0.76	1.58	2.17
90	832.1	725.3	2.59	0.163	0.37	1.07	1.43	1.65	2.04
110	786.6	704.1	5.17	0.160	0.28	1.13	2.66	1.72	1.89
130	734.4	678.7	9.25	0.159	0.21	1.18	4.30	1.78	1.75

HEPTANE

$^\circ$C	kJ/kg	kg/m^3	kg/m^3	W/m$^\circ$C	cP	cPx10^2	Bar	kJ/kg$^\circ$C	N/mx10^2
-20	384.0	715.5	0.01	0.143	0.69	0.57	0.01	0.83	2.42
0	372.6	699.0	0.17	0.141	0.53	0.60	0.02	0.87	2.21
20	362.2	683.0	0.49	0.140	0.43	0.63	0.08	0.92	2.01
40	351.8	667.0	0.97	0.139	0.34	0.66	0.20	0.97	1.81
60	341.5	649.0	1.45	0.137	0.29	0.70	0.32	1.02	1.62
80	331.2	631.0	2.31	0.135	0.24	0.74	0.62	1.05	1.43
100	319.6	612.0	3.71	0.133	0.21	0.77	1.10	1.09	1.28
120	305.0	592.0	6.08	0.132	0.18	0.82	1.85	1.16	1.10

WATER

Temp.	Latent Heat	Liquid Density	Vapour Density	Liquid Thermal Conductivity	Liquid Viscos.	Vapour Viscos.	Vapour Press.	Vapour Specific Heat	Liquid Surface Tension
$^{\circ}$C	kJ/kg	kg/m^3	kg/m^3	W/m$^{\circ}$C	cP	cPx10^2	Bar	kJ/kg$^{\circ}$C	N/mx10^2
20	2448	998.2	0.01	0.612	1.00	0.96	0.02	1.85	7.40
40	2402	992.3	0.05	0.630	0.65	1.04	0.07	1.86	6.96
60	2359	983.0	0.14	0.649	0.47	1.12	0.20	1.87	6.62
80	2309	972.0	0.29	0.668	0.36	1.19	0.47	1.88	6.26
100	2258	958.0	0.60	0.680	0.28	1.27	1.01	1.88	5.89
120	2200	945.0	1.12	0.682	0.23	1.34	2.02	1.89	5.50
140	2139	928.0	1.99	0.683	0.20	1.41	3.90	1.90	5.06
160	2074	909.0	3.27	0.679	0.17	1.49	6.44	1.91	4.66
180	2003	888.0	5.16	0.669	0.15	1.57	10.04	1.92	4.29
200	1967	865.0	7.87	0.659	0.14	1.65	16.19	1.93	3.89

FLUTEC PP9

$^{\circ}$C	kJ/kg	kg/m^3	kg/m^3	W/m$^{\circ}$C	cP	cPx10	Bar	kJ/kg$^{\circ}$C	N/mx10^2
-30	103.0	2098	0.01	0.060	5.77	0.82	0.00	0.80	2.36
0	98.4	2029	0.01	0.059	3.31	0.90	0.00	0.87	2.08
30	94.5	1960	0.12	0.057	1.48	1.06	0.01	0.94	1.80
60	90.2	1891	0.61	0.056	0.94	1.18	0.03	1.02	1.52
90	86.1	1822	1.93	0.054	0.65	1.21	0.12	1.09	1.24
120	83.0	1753	4.52	0.053	0.49	1.23	0.28	1.15	0.95
150	77.4	1685	11.81	0.052	0.38	1.26	0.61	1.23	0.67
180	70.8	1604	25.13	0.051	0.30	1.33	1.58	1.30	0.40
225	59.4	1455	63.27	0.049	0.21	1.44	4.21	1.41	0.01

THERMEX
(DIPHENYL – DIPHENYL OXIDE EUTECTIC)

Temp.	Latent Heat	Liquid Density	Vapour Density	Liquid Thermal Conductivity	Liquid Viscos.	Vapour Viscos.	Vapour Press.	Vapour Specific Heat	Liquid Surface Tension
$^{\circ}C$	kJ/kg	kg/m^3	kg/m^3	W/m$^{\circ}$C	cP	cPx10	Bar	kJ/kg$^{\circ}$C	N/mx10^2
100	354.0	992.0	0.03	0.131	0.97	0.67	0.01	1.34	3.50
150	338.0	951.0	0.22	0.125	0.57	0.78	0.05	1.51	3.00
200	321.0	905.0	0.94	0.119	0.39	0.89	0.25	1.67	2.50
250	301.0	858.0	3.60	0.113	0.27	1.00	0.88	1.81	2.00
300	278.0	809.0	8.74	0.106	0.20	1.12	2.43	1.95	1.50
350	251.0	755.0	19.37	0.099	0.15	1.23	5.55	2.03	1.00
400	219.0	691.0	41.89	0.093	0.12	1.34	10.90	2.11	0.50
450	185.0	625.0	81.00	0.086	0.10	1.45	19.00	2.19	0.03

MERCURY

$^{\circ}C$	kJ/kg	kg/m^3	kg/m^3	W/m$^{\circ}$C	cP	cPx10	Bar	kJ/kg$^{\circ}$Cx10	N/mx10
150	308.8	13230	0.01	9.99	1.09	0.39	0.01	1.04	4.45
250	303.8	12995	0.60	11.23	0.96	0.48	0.18	1.04	4.15
300	301.3	12880	1.73	11.73	0.93	0.53	0.44	1.04	4.00
350	298.9	12763	4.45	12.18	0.89	0.61	1.16	1.04	3.82
400	296.3	12656	8.75	12.58	0.86	0.66	2.42	1.04	3.74
450	293.8	12508	16.80	12.96	0.83	0.70	4.92	1.04	3.61
500	291.3	12308	28.60	13.31	0.80	0.75	8.86	1.04	3.41
550	288.8	12154	44.92	13.62	0.79	0.81	15.03	1.04	3.25
600	286.3	12054	65.75	13.87	0.78	0.87	23.77	1.04	3.15
650	283.5	11962	94.39	14.15	0.78	0.95	34.95	1.04	3.03
750	277.0	11800	170.00	14.80	0.77	1.10	63.00	1.04	2.75

CAESIUM

Temp.	Latent Heat	Liquid Density	Vapour Density	Liquid Thermal Conductivity	Liquid Viscos.	Vapour Viscos.	Vapour Press.	Vapour Specific Heat	Liquid Surface Tension
°C	kJ/kg	kg/m³	kg/m³x10²	W/m°C	cP	cPx10²	Bar	kJ/kg°Cx10	N/mx10²
375	530.4	1740	0.01	20.76	0.25	2.20	0.02	1.56	5.81
425	520.4	1730	0.01	20.51	0.23	2.30	0.04	1.56	5.61
475	515.2	1720	0.02	20.02	0.22	2.40	0.09	1.56	5.36
525	510.2	1710	0.03	19.52	0.20	2.50	0.16	1.56	5.11
575	502.8	1700	0.07	18.83	0.19	2.55	0.36	1.56	4.81
625	495.3	1690	0.10	18.13	0.18	2.60	0.57	1.56	4.51
675	490.2	1680	0.18	17.48	0.17	2.67	1.04	1.56	4.21
725	485.2	1670	0.26	16.83	0.17	2.75	1.52	1.56	3.91
775	477.8	1655	0.40	16.18	0.16	2.82	2.46	1.56	3.66
825	470.3	1640	0.55	15.53	0.16	2.90	3.41	1.56	3.41

POTASSIUM

°C	kJ/kg	kg/m³	kg/m³	W/m°C	cP	cPx10	Bar	kJ/kg°Cx10	N/mx10²
350	2093	763.1	0.002	51.08	0.21	0.15	0.01	5.32	9.50
400	2078	748.1	0.006	49.08	0.19	0.16	0.01	5.32	9.04
450	2060	735.4	0.015	47.08	0.18	0.16	0.02	5.32	8.69
500	2040	725.4	0.031	45.08	0.17	0.17	0.05	5.32	8.44
550	2020	715.4	0.062	43.31	0.15	0.17	0.10	5.32	8.16
600	2000	705.4	0.111	41.81	0.14	0.18	0.19	5.32	7.86
650	1980	695.4	0.193	40.08	0.13	0.19	0.35	5.32	7.51
700	1960	685.4	0.314	38.08	0.12	0.19	0.61	5.32	7.12
750	1938	675.4	0.486	36.31	0.12	0.20	0.99	5.32	6.72
800	1913	665.4	0.716	34.81	0.11	0.20	1.55	5.32	6.32
850	1883	653.1	1.054	33.31	0.10	0.21	2.34	5.32	5.92

SODIUM

Temp.	Latent Heat	Liquid Density	Vapour Density	Liquid Thermal Conductivity	Liquid Viscos.	Vapour Viscos.	Vapour Press.	Vapour Specific Heat	Liquid Surface Tension
°C	kJ/kg	kg/m³	kg/m³	W/m°C	cP	cPx10	Bar	kJ/kg°Cx10	N/mx10
500	4370	828.1	0.003	70.08	0.24	0.18	0.01	9.04	1.51
600	4243	805.4	0.013	64.62	0.21	0.19	0.04	9.04	1.42
700	4090	763.5	0.050	60.81	0.19	0.20	0.15	9.04	1.33
800	3977	757.3	0.134	57.81	0.18	0.22	0.47	9.04	1.23
900	3913	745.4	0.306	53.35	0.17	0.23	1.25	9.04	1.13
1000	3827	725.4	0.667	49.08	0.16	0.24	2.81	9.04	1.04
1100	3690	690.8	1.306	45.08	0.16	0.25	5.49	9.04	0.95
1200	3577	669.0	2.303	41.08	0.15	0.26	9.59	9.04	0.86
1300	3477	654.0	3.622	37.08	0.15	0.27	15.91	9.04	0.77

LITHIUM

°C	kJ/kg	kg/m³	kg/m³	W/m°C	cP	cPx10²	Bar	kJ/kg°C	N/mx10
1030	20500	450	0.005	67	0.24	1.67	0.07	0.532	2.90
1130	20100	440	0.013	69	0.24	1.74	0.17	0.532	2.85
1230	20000	430	0.028	70	0.23	1.83	0.45	0.532	2.75
1330	19700	420	0.057	69	0.23	1.91	0.96	0.532	2.60
1430	19200	410	0.108	68	0.23	2.00	1.85	0.532	2.40
1530	18900	405	0.193	65	0.23	2.10	3.30	0.532	2.25
1630	18500	400	0.340	62	0.23	2.17	5.30	0.532	2.10
1730	18200	398	0.490	59	0.23	2.26	8.90	0.532	2.05

Appendix 2
Thermal Conductivity of Heat Pipe Container and Wick Materials

Material	Thermal Conductivity (W/m$^{\circ}$C)
Aluminium	205
Brass	113
Copper (0 - 100°C)	394
Glass	0.75
Nickel (0 - 100°C)	88
Mild Steel	45
Stainless Steel (Type 304)	17.3
Teflon	0.17

Appendix 3
The Navier-Stokes Equation

The Equation of motion of a uniform, viscous fluid, of constant viscosity, may be written:

$$\rho \frac{Dv}{Dt} = - \text{grad } P + \mu \nabla^2 v + \frac{\mu}{3} \text{ grad div } v + \rho F \qquad \dots \text{A.3.1}$$

where $\frac{D}{Dt}$ is the 'substantial time derivative'

which may be written in rectangular coordinates

$$\frac{D}{Dt} = \frac{\partial}{\partial t} + v_x \frac{\partial}{\partial x} + v_y \frac{\partial}{\partial y} + v_z \frac{\partial}{\partial z}$$

- grad P represents the force due to pressure,
$\mu \nabla^2 v + \frac{\mu}{3} \text{ grad div } v$ the force due to viscous effects and ρF the body forces.

If only gravity and body forces are present this reduces to ρg.   If, as is usually the case, we can assume incompressible flow, Equation A.3.1 reduces to:

$$\rho \frac{Dv}{Dt} = - \text{grad } P + \mu \nabla^2 v + \rho F \qquad \dots \text{A.3.2}$$

Equation A.3.2 is the Navier Stokes Equation.

If the fluid is inviscid $\mu = 0$ and it reduces to the Euler Equation:

$$\rho \frac{Dv}{Dt} = - \text{grad } P + \rho g$$

The components of the Navier Stokes Equation are given below for rectangular and cylindrical coordinates.

A.3.1 Navier Stokes Equation. Rectangular coordinates x, y, z.

x component
$$\rho\left[\frac{\partial v_x}{\partial t} + v_x\frac{\partial v_x}{\partial x} + v_y\frac{\partial v_x}{\partial y} + v_z\frac{\partial v_x}{\partial z}\right] = -\frac{\partial P}{\partial x}$$

$$+ \mu\left[\frac{\partial^2 v_x}{\partial x^2} + \frac{\partial^2 v_y}{\partial y^2} + \frac{\partial^2 v_z}{\partial z^2}\right] + \rho g_x$$

y component
$$\rho\left[\frac{\partial v_y}{\partial t} + v_x\frac{\partial v_y}{\partial x} + v_y\frac{\partial v_y}{\partial y} + v_z\frac{\partial v_y}{\partial z}\right] = -\frac{\partial P}{\partial y}$$

$$+ \mu\left[\frac{\partial^2 v_y}{\partial x^2} + \frac{\partial^2 v_y}{\partial y^2} + \frac{\partial^2 v_y}{\partial z^2}\right] + \rho g_y$$

z component
$$\rho\left[\frac{\partial v_z}{\partial t} + v_x\frac{\partial v_z}{\partial x} + v_y\frac{\partial v_z}{\partial y} + \partial v_z\frac{\partial v_z}{\partial z}\right] = -\frac{\partial P}{\partial z}$$

$$+ \mu\left[\frac{\partial^2 v_z}{\partial x^2} + \frac{\partial^2 v_z}{\partial y^2} + \frac{\partial^2 v_z}{\partial z^2}\right] + \rho g_z$$

Cylindrical coordinates r, θ, z.

r component
$$\rho\left[\frac{\partial v_r}{\partial t} + v_r\frac{\partial v_r}{\partial r} + \frac{v_\theta}{r}\frac{\partial v_r}{\partial \theta} - \frac{v_\theta^2}{r} + v_z\frac{\partial v_r}{\partial z}\right] = -\frac{\partial P}{\partial r}$$

$$+ \mu\left[\frac{\partial}{\partial r}\left[\frac{1}{r}\frac{\partial}{\partial r}(rv_r)\right] + \frac{1}{r^2}\frac{\partial^2 v_r}{\partial \theta^2} - \frac{2}{r^2}\frac{\partial v_\theta}{\partial \theta} + \frac{\partial^2 v_r}{\partial z^2}\right] + \rho g_r$$

θ component

$$\rho\left[\frac{\partial v_\theta}{\partial t} + v_r\frac{\partial v_\theta}{\partial r} + \frac{v_\theta}{r}\frac{\partial v_\theta}{\partial \theta} + \frac{v_r v_\theta}{r} + v_z\frac{\partial v_\theta}{\partial z}\right] = -\frac{1}{r}\frac{\partial P}{\partial \theta}$$

$$+ \mu\left[\frac{\partial}{\partial r}\left[\frac{1}{r}\frac{\partial}{\partial r}(rv_\theta)\right] + \frac{1}{r^2}\frac{\partial^2 v_\theta}{\partial \theta^2} + \frac{2}{r^2}\frac{\partial v_r}{\partial \theta} + \frac{\partial^2 v_\theta}{\partial z^2}\right] + \rho g_\theta$$

z component

$$\rho\left[\frac{\partial v_z}{\partial t} + v_r\frac{\partial v_z}{\partial r} + \frac{v_\theta}{r}\frac{\partial v_z}{\partial \theta} + v_z\frac{\partial v_z}{\partial z}\right] = -\frac{\partial P}{\partial z}$$

$$+ \mu\left[\frac{1}{r}\frac{\partial}{\partial r}\left[\frac{r\partial v_z}{\partial z}\right] + \frac{1}{r^2}\frac{\partial^2 v_z}{\partial \theta^2} + \frac{\partial^2 v_z}{\partial \theta^2} + \frac{\partial^2 v_z}{\partial z^2}\right] + \rho g_z$$

The solution of these equations presents a formidable mathematical problem. They can usually be simplified, for example apart from start up it is possible to consider steady state conditions when $\frac{\partial}{\partial t} = 0$. Most heat pipe treatments have assumed cylindrical symmetry when $\frac{\partial}{\partial \theta} = 0 = v_\theta$. The v_θ component is however required in the case of the rotating heat pipe.

A.3.2 The Equation of Continuity.   The equation of continuity for a fluid is:

$$\frac{\partial \rho}{\partial t} = -(\nabla\rho.v)$$

It may also be expressed in terms of the substantial time derivative

$$\frac{D\rho}{Dt} = -\rho(\nabla.v) \qquad\qquad ... A.3.3$$

For an incompressible fluid

$$\nabla.v = 0$$

In rectangular coordinates the continuity equation is written:

$$\frac{\partial \rho}{\partial t} + \frac{\partial}{\partial x}(\rho v_x) + \frac{\partial}{\partial y}(\rho v_y) + \frac{\partial}{\partial z}(\rho v_z) = 0$$

and in cylindrical coordinates

$$\frac{\partial \rho}{\partial t} + \frac{1}{r}\frac{\partial}{\partial r}(\rho r v_r) + \frac{1}{r}\frac{\partial}{\partial \theta}(\rho v_\theta) + \frac{\partial}{\partial z}(\rho v_z) = 0$$

REFERENCES

A.3.1   Kay, J.M.   An introduction to fluid mechanics and heat transfer.
        2nd Ed.   Pub. Cambridge University Press, 1963.

A.3.2   Bird, R.B., Stewart, W.E. and Lightfoot, E.N.   Transport Phenomena.
        Pub. Wiley & Sons, Inc. N.Y. 1960.

Appendix 4
Suppliers of Materials of use
in Heat Pipe Manufacture

Wick Materials

Foam Metals

Hogen Industries
37645 Vine Street,
Willoughby,
Ohio 44094,
USA.

Astro Met Associates,
95 Barron Drive,
Cincinnati,
Ohio 45215,
USA.

Gould Inc.,
Gould Laboratories,
540 East 105th Street,
Cleveland,
Ohio 44108,
USA.

Refrasil Sleeving

The Chemical & Insulating Co. Ltd.,
Darlington,
UK.

Sintered Materials

Sintered Products Ltd.,
Hamilton Road,
Sutton-in-Ashfield,
Notts. NG17 5LL,
UK.

British Metal Sinterings Association,
C/O Peat, Marwick, Mitchell Co. Ltd.,
Windsor House,
Temple Row,
Birmingham 2,
UK.

Metal Powder

Goodfellow Metals Ltd.,
Cambridge Science Park,
Milton Road,
Cambridge CB4 4DS,
UK.
(Small quantities).

Ronald Britton Ltd.,
Woodbine Street East,
Rochdale,
Lancs. OL16 5JE,
UK.
(Commercial quantities).

Wire Mesh

Begg, Cousland & Co. Ltd.,
Springfield Wire Works,
636, Springfield Road,
Glasgow G40 3HS,
Scotland.

N. Greenings Ltd.,
Brittania Works,
Warrington WA5 5JX,
Lancashire,
UK.

Associated Perforators & Weavers,
Church Street,
Warrington,
Lancashire,
UK.

R. Cadisch & Sons,
Arcadia Avenue,
Finchley,
London N3 2JZ.

F.W. Potter & Soar Ltd.,
Beaumont Road,
Banbury,
Oxon OX16 7SD,
UK.

United Wire Ltd.,
Myddleton Road,
London N22 4NH,
UK.

Cambridge Wire Cloth,
P.O. Box 399,
Cambridge,
Maryland 21613,
USA.

Michigan Wire Cloth Co. Inc.,
2100 Howard Street,
Detroit,
Michigan, 48216,
USA.

Metal Felts   Astro Met Associates. Inc.,
                                   95 Barron Drive,
                                   Cincinnati,
                                   Ohio 45215,
                                   USA.

Greaseless High Vacuum Valves (glass)

                                   J. Young (Scientific Glassware) Ltd.,
                                   11 Colville Road,
                                   Acton,
                                   London W3 8BS,
                                   UK.

Metal Valves   Techmation Ltd.,
                                   58 Edgware Way,
                                   Edgware,
                                   Middx. HA8 8JP,
                                   UK.

                                   Hoke International Ltd.,
                                   Brookhill Road,
                                   Barnet,
                                   Hertfordshire EN4 8AU,
                                   UK.

Stainless Steel Tubes

                                   Oxford Instruments,
                                   Osney Mead,
                                   Oxford OX2 ODX,
                                   UK.

                                   Fine Tubes Ltd.,
                                   Estover Works,
                                   Crownhill,
                                   Plymouth PL6 7LG,
                                   UK.

Porous Metal Laminates

                                   Gould Inc.,
                                   540 East 105th Street,
                                   Cleveland,
                                   Ohio 44108,
                                   USA.

Appendix 5
Commercial Heat Pipe Manufacturers

Isothermics Inc.,
P.O. Box 86,
Augusta,
New Jersey 07822.
USA.

Hughes Aircraft Company,
Electron Dynamics Division,
3100 West Lomita Boulevard,
Torrance,
California 90509,
USA.

Thermo Electron Corp.,
85 First Avenue,
Waltham,
Massachusetts 02154,
USA.

Power Technology Corp.,
105 Enterprise Drive,
Ann Arbor,
Michigan 48103,
USA.

Q-Dot Corp.,
151 Regal Row,
Suite 120,
Dallas,
Texas 75247,
USA.

McDonnell Douglas Corp.,
Donald W. Douglas Labs.,
2955 George Washington Way,
Richland,
Washington 99352,
USA.

Noren Products Inc.,
846 Blandford Blvd.,
Redwood City,
California 94062,
USA.

Energy Conversion Systems,
623 Wyoming SE,
Albuquerque,
New Mexico 87112,
USA.

Heat Pipe Corp. of America,
141 Park Place,
Watchung,
New Jersey 07060,
USA.

Dynatherm Corp.,
Baltimore,
Md. 21204,
USA.

Redpoint Associates,
Lynton Road,
Cheyney Manor,
Swindon,
Wiltshire,
UK.

Solek Ltd.,
16 Hollybush Lane,
Sevenoaks,
Kent,
UK.

International Research & Development
                        Co. Ltd.,
Fossway,
Newcastle-upon-Tyne NE6 2YD,
UK.

Scurrah Hytech,
Wellington Arms Hotel,
Stratfield Turgis,
Basingstoke,
Hants. RG27 0AS,
UK.

Isoterix Ltd.,
1 Bank House,
Balcombe Road,Horley,
Surrey,
UK.

Appendix 6
Bibliography on Heat Pipe Applications

Abu-Romia, M.M. and Bhatia, B. Measurement of stagnation point heat transfer from plasma torch using heat pipe calorimetry. AIAA Paper 71-81. 9th Aerospace Sciences Meeting, New York. Jan 1971, AIAA, 1971.

Anand, D.K. et al. Heat pipe application for spacecraft thermal control, JHU Tech. Memo. APL-TG-922. Johns Hopkins University, Applied Physics Lab., August 1967. AD 662 241.

Anderson, J.L. and Lantz, E. A nuclear thermionic space power concept using rod control and heat pipes. NASA TM-X-52446. Lewis Research Center, 1968.

Ando, M. Application of heat pipes to nuclear steelmaking. Nuclear Engng. Int., pp 38-39, June 1976.

Anon. Improving fuel vaporization. Automotive Engng., Vol. 84, No. 6, pp 37-43, June 1976.

Asselman, G.A.A. Thermal energy storage unit based on lithium fluoride. Energy Conversion, Vol. 16, No. 1-2, 1976.

Asselman, G.A.A. and Green, D.B. Heat Pipes, II - Applications. Philips Tech. Review, Vol. 33, No. 4, pp 104 - 113, 1973.

Basiulis, A. Uni-directional heat pipes to control TWT temperature in synchronous orbit. Proc. Thermodynamics & Thermophysics of Space Symp. Palo Alto, Calif., pp 165 - 173, 1970.

Basiulis, A. and Dixon, J.C. Heat pipe design for electron tube cooling. ASME Paper 69-HT-25, 1969.

Basiulis, A. and Filler, M. Characteristics of six novel heat pipes for thermal control applications. ASME Paper 71-Av-29, 1971.

Basiulis, A. and Hummel, T.A. The application of heat pipe techniques to electronic component cooling. ASME Paper 72-WA/HT-42, 1972.

Basiulis, A. and Johnson, J.H. High temperature heat pipes for energy conservation. Proc. 10th Intersoc. Energy Conversion and Engng. Conf., Newark, August 1975.

Berger, M.E. and Kelly, W.H. Application of heat pipes to the ATS-F Spacecraft. ASME Paper 73-ENAs-46, 1973.

Bienert, W.B. Heat pipes for solar energy collectors. 1st Int. Heat Pipe Conf. Paper 12-1, Stuttgart, Oct. 1973.

Bienert, W.B. and Kroliczek, E. Experimental high performance heat pipes for the OAO-C spacecraft. ASME Paper 71-Av-26, 1971.

Birnbeier, H. et al. Heat pipe cooling of semi-conductor power devices. Proc. 10th Intersoc. Energy Conversion & Engng. Conf., Newark, August 1975.

Bohdansky, J. and Schins, H.E.J. New method for vapor-pressure measurements at high temperature and high pressure. J. Appl. Phys., Vol. 36, No. 11, pp. 3683 - 3684, 1965.

Breitweiser, R. Use of heat pipes for electrical isolation. IEEE 1970. Thermionic Conversion Specialist Conf. Oct. 1970, pp 185 - 190.

Brost, O. et al. Industrial applications of alkali-metal heat pipes. 1st Int. Heat Pipe Conf. Paper 11-3, Stuttgart, Oct. 1973.

Brost, O. and Muenzel, W.D. Heat pipe applications development in Europe. Proc. 10th Intersoc. Energy Conversion & Engng. Conf., Newark, August 1975.

Brost, O. and Schubert, K.P. Development of alkali-metal heat pipes as thermal switches. 1st Int. Heat Pipe Conf. Paper 12-2, Stuttgart, 1973.

Brown, A. and Morris, K.J. Heat pipes for domestic use. 1st Int. Heat Pipe Conf. Paper 11-2. Stuttgart, 1973.

Busse, C.A., Caron, R. and Cappelletti, C. Prototypes of heat pipe thermionic converters for space reactors. IEE, 1st Int. Conf. on Thermionic Elect. Power Generation, London, 1965.

Calimbas, A.T. and Hulett, R.H. An avionic heat pipe. ASME Paper 69-HT-16, 1969.

Conway, E.C. and Kelley, M.J. A continuous heat pipe for spacecraft thermal control. Proc. Annual Aviation & Space Conf., Beverley Hills, California, June, 1968, pp 655 - 658, ASME 1968.

Corman, J.C. and McLaughlin, M.H. Thermal design of heat pipe cooled a.c. motor, ASME Paper 71-WA/HT-14, 1971.

Dean, D.J. An integral heat pipe package for microelectronic circuits. Proc. 2nd Int. Heat Pipe Conference, Bologna, 1976.

Deverall, J.E. Heat pipe thermal control of irradiation capsules. 1st Int. Heat Pipe Conf. Paper 8-2, Stuttgart, Oct. 1973.

Dunn, P.D. and Rice, G. Reactor fuel test rig using heat pipe control. 1st Int. Heat Pipe Conf. Paper 8-3, Stuttgart, Oct. 1973.

Dutcher, C.H. and Burke, M.R. Heat pipes - a cool way to cool circuitry. Electronics, Vol. 34, No. 4, pp 94 - 100, 1970.

Eddleston, B.N.F. and Hecks, K. Application of heat pipes to the thermal control of advanced communications spacecraft. 1st Int. Heat Pipe Conf., Paper 9 - 4, Stuttgart, Oct. 1973.

Edwards, J.P. Liquid and vapour cooling systems for gas turbines. ARC-CP-1127.
London: HMSO, 1970.

Feldman, K.T. and Whiting, G.H. Applications of the heat pipe. Mech. Engng.
Vol. 90, pp 48 - 53. Nov. 1968.

Ferrara, A. and Brinkman, P. Applying heat pipes to avoid the preferential
freezing of highway bridge decks. ASME Paper 76-ENAs-25, 1976.

Finlay, I.C. and Gee, D.G. Performance of a prototype isothermal oven for use
with an 'X'-band microwave noise standard. Proc. 2nd Int. Heat Pipe Conference,
Bologna, 1976.

Finlay, I.C. and Green, D.B. Heat pipes and their instrument applications.
J. Physics E: Scientific Instruments, Vol. 9, pp 1026 -1035, 1976.

Fitton, G.L. Taking out the heat. Electronic Engineering, Vol. 45, No. 550,
Dec. 1973.

Gavrilov, A.F. Design of an air heater with an intermediate heat carrier.
Teploenergetika, Vol. 13, No. 8, pp. 92 - 93, 1966. Thermal Engineering, Vol.
13, No. 8, pp 121 - 124, 1966.

Gerrels, E.E. and Larson, J.W. Brayton cycle vapour chamber (heat pipe) radiator study. NASA CR-1677, General Electric Company, Feb. 1971.

Green, D.B. A thermal conductivity apparatus employing heat pipes. Philips
Research Laboratory Report, Eindhoven, 1972.

Groll, M. et al. Industrial applications of low temperature heat pipes. 1st
Int. Heat Pipe Conf. Paper 11-1, Stuttgart, Oct. 1973.

Groll, M. et al. An electrical feedback controlled high capability variable
conductance heat pipe for satellite application. 1st Int. Heat Pipe Conf.
Paper 8-1, Stuttgart, Oct. 1973.

Guenard, P. High power linear beam tube devices. J. Microwave Power, Vol. 5,
pp 261 - 267, Dec. 1970.

Harwell, W. et al. Orbiting astronomical laboratory heat pipe flight performance data. AIAA Paper 73-758, 1973.

Harbaugh, W.E. and Eastman, G.Y. Applying heat pipes to thermal problems.
Heat Piping and Air Condit. Vol. 42, No. 10, pp 92 - 96, Oct. 1970.

Hassan, H. and Accensi, A. Spacecraft applications of low temperature heat
pipes, 1st Int. Heat Pipe Conf., Paper 9-1, Stuttgart, Oct. 1973.

Hinderman, J.D. et al. An ATS-E solar cell space radiator utilising heat
pipes. AIAA Paper 69-630, 1969.

Hoppe, U. et al. Development of heat pipe radiator elements. 1st Int. Heat
Pipe Conf., Paper 10-3, Stuttgart, Oct. 1973.

Johns Hopkins University, Applied Physics Lab., The Geos-II heat pipe system and its performance in test and in orbit. NASA CR-94585, JHU 52P-3-25, April 1968.

Kirkpatrick, J.P. and Marcus, B.D. A variable conductance heat pipe flight experiment. AIAA Paper 71-411, 1971.

Kopf, L. A low temperature heat pipe used as a thermal switch. Rev. Scient. Instrum. Vol. 42, No. 12, pp 1764 - 1765, Dec. 1971.

Kroliczek, E. Heat pipe heat rejection system for electrical batteries. NASA Report CR-144763, Dynatherm Corporation, 1976.

Larkin, B.S. and Johnston, G.H. An experimental field study of the use of 2-phase thermosiphons for the preservation of permafrost. National Research Council, Canada. Paper presented at Annual Congress of the Engineering Institute of Canada, Montreal, 2nd Oct. 1973.

Larkin, B.S. A computer cooling application for thermosiphons. Engineering Journal (Canada) pp 29 - 33. Jan. 1973.

Larkin, B.S. et al. A thermosiphon heat exchanger for use in animal shelters. Paper 74-210, Canadian Society of Agricultural Engineers, 1974.

Leeth, G.G. Energy transmission systems. J. Hydrogen Energy, Vol. 1, pp 49 - 53, Jan. 1976.

Lidbury, J.A. A helium heat pipe. Rutherford High Energy Laboratory, NDG Report 72-11, April 1972.

Lindsey, R. and Wilson, J. Heat pipe vaporisation of gasoline - vapipe. 1st Symposium on Low Pollution Power Systems Development, Ann Arbor, Oct. 14 - 19, 1973.

Losch, H.E. and Pawlowski, P.H. Heat pipe and phase changing material (PCM) sounding rocket experiment. AIAA Paper 73-759, 1973.

Martinez, I. et al. Development and test of a heat pipe radiative plate for space applications. 1st Int. Heat Pipe Conf., Stuttgart, 1973.

McIntosh, R. and Ollendorf, S. The International Heat Pipe Experiment. Proc. 2nd Int. Heat Pipe Conference, Bologna, 1976. (Spacecraft applications).

Milleron, N. and Wolgast, R. Cryopumping the omnitron ultra-vacuum system using heat pipes and metallic conductors. IEEE Trans. Nuclear Sci., Vol. NS-16, Pt. 3, pp 941 - 944. June 1969.

NBS Technical News Bulletin. Heat pipe oven generated homogeneous metal vapours. NBS Tech. News Bull., Vol. 53, No. 8, pp 172 - 173, 1969.

Oslejsek, O. and Polasek, F. Cooling of electrical machines by heat pipes. Proc. 2nd Int. Heat Pipe Conference, Bologna, 1976.

Pitts, J.H. and Walter, C.E. Conceptual design of a 10 MWe nuclear Rankine system for space power. J. Spacecraft & Rockets, Vol. 7, No. 3, pp 259 - 265, 1970.

Puthoff, R.L. and Silverstein, C.C. Application of heat pipes to a nuclear aircraft propulsion system. AIAA Paper 70-662, 1970.

Reay, D.A. The heat pipe exchanger: a technique for waste heat recovery. Heating and Ventilating Engineer, Vol. 50, No. 593, Jan. 1977.

Reay, D.A. Heat pipes: Applications in zinc and aluminium diecasting. Die-casting and Metal Moulding, Vol. 6, No. 12, Nov/Dec. 1975.

Reay, D.A. Heat pipe cooling offers many advantages. Electronics Engng. Vol. 44, No. 8, pp 35 – 37, 1972.

Reay, D.A. The heat pipe: Its development and its aerospace applications. The Aeronautical Journal, R.Ae.S., Vol. 78, No. 765, pp 414 – 423, Sept. 1974.

Reay, D.A. and Summerbell, D. The development of heat pipes for satellites. 1st Int. Heat Pipe Conf., Paper 9.2, Stuttgart, Oct. 1973.

Rice, G. and Hammerton, J.C. Investigation into possible means of fuel saving in steel slab reheat furnaces. Report No. 4. Reading University, Department of Applied Physical Sciences, Reading, Berks. Dec. 1968.

Roberts, C.C. Designing heat pipe heat sinks. Proc. 10th AIAA Thermophysics Conf., Denver, Colorado, May 1975.

Roukis, J. et al. Heat pipe applications to space vehicles. AIAA Paper 71 – 410, 1971.

Rousar, D.C. Heat pipe cooled thrust chambers for space storable propellants. AIAA Paper 70-942, 1970.

Savage, C.J. Heat pipes and vapour chambers for satellite thermal balance. RAE TR-69125, Farnborough, Hants. Royal Aircraft Establishment, June 1969.

Shefsiek, P.K. and Ernst, D.M. Heat pipe development for thermionic converter applications. 4th Intersoc. Energy Conversion Conf., Washington D.C., 22 – 26 Sept. 1969, pp 879 – 887. AIChE, 1969.

Shlosinger, A.P. Heat pipe devices for space suit temperature control. NASA CR-1400, TRW Systems Group, Oct. 1969.

Shukin, V.K. et al. A gas turbine engine regenerator employing heat pipes. Aviatsionnaia Tekhnika, Vol. 19, No. 1, pp 172 – 175, 1976.

Silverstein, C.C. Heat pipe gas turbine regenerators. ASME Paper 68-WA/GT-7, 1968.

Silverstein, C.C. A feasibility study of heat pipe cooled leading edges for hypersonic cruise aircraft. NASA CR-1857, Nov. 1971.

Tani, T. et al. A terrestrial solar energy power system. Solar Energy, Vol. 18, No. 4, pp 281 – 285, 1976.

Thurman, J.L. and Ingram, E.H. Application of heat pipes to reduce cryogenic boil-off in space. J. Spacecraft & Rockets, Vol. 6, No. 3, pp 319 – 321, March 1969.

Vidal, C.R. and Haller, F.B. Heat pipe oven applications. I. Isothermal heater of well defined temperature. II. Production of metal vapour-gas mixtures. Rev. Scient. Instrum. Vol. 42, No. 12, pp 1779 - 1784, Dec. 1971.

Waters, E.D. Arctic tundra kept frozen by heat pipes. The Oil and Gas Journal (US), Aug. 26, 1974.

Weismantel, G.E. Alaska pipeline spinoffs. Chemical Engng, Vol. 81, No. 6, pp 42 - 44, March 18, 1974.

Wilson, J.L. Developments in heat transfer of interest to the heating and ventilating engineer. J. Instn. Heat. Vent. Engnrs. Vol. 39, No. 9, pp 123 - 127, 1971.

Wright, J.P. and Pence, W.R. Development of a cryogenic heat pipe radiator for a detector cooling system. ASME Paper 73-ENAs-47, 1973.

Wright, J.P. and Trucks, H. Thermal control systems for low temperature shuttle payloads. ASME Paper 76-ENAs-65, 1976.

Zimmermann, P. and Pruschek, R. Principles and industrial applications of heat pipes (in German). DECHEMA. Monogr. Vol. 65, (Nos 1168 - 1192), pp 67 - 84, 1970.

Appendix 7
Heat Pipe Patents

The following list of heat pipe patents filed in the United Kingdom and the United States of America includes significant and interesting developments and applications. This list is not exhaustive, but includes patents published during 1976. The seven-figure number is the patent number and the date given is the patent application date.

A7.1 The Heat Pipe Concept

2279548 (US)	11 June 1938. Liquid Vaporizing Tube. This patent describes a tube incorporating capillary grooves to aid liquid distribution and hence vaporization in boilers.
2350348 (US)	21 Dec. 1942. Gaugler's first patent describing the heat pipe.
3229759 (US)	2 Dec. 1963. Grover's heat pipe patent. Seven other patents were cited as a result of the Patent Office search, including Gaugler's, but the patent was accepted. Filed in the UK on 25 Nov. 1964 (1027719).

A7.2 Wick Geometries

3528494 (US)	7 Nov. 1966. W.J. Levedahl describes a heat pipe having axial grooves in the wall. It is directed particularly at units using a low conductivity working fluid.
1118468 (UK)	16 Nov. 1965. Euratom patent also describing longitudinal grooves, restricted to particular geometries.
1275946 (UK)	25 Aug. 1969. Junkers describe a large number of arterial types of wick (approximately 20 in all) suitable for heat pipes for space use.
3786861 (US)	12 April, 1971. Battelle Institute patent on a wick of parallel longitudinal capillary channels. Cross-sectional area of each channel is 10^{-4} to 10^{-1} mm^2 and the wick is chamfered at the evaporator and condenser to permit release of and access for liquid.
3840069 (US)	20 April, 1972. Brown, Boveri & Cie A.G. This patent describes a sintered wick structure, in which the pore size is selected prior to manufacture to include both coarse and fine grades.

3746081   17 July 1973. General Electric Company patent in which
(US)   radial holes in the evaporator wick are claimed to mini-
          mise radial resistance to vapour flow. Claims that the
          heat transport capability of a water heat pipe could be
          doubled using this technique. (See also subsequent US
          patents 3828849 and 3955619, published in 1974 and 1976
          respectively).

3893293   1 July 1975 (issued). Perkin-Elmer Corporation patent for
(US)   a high capability lobar arterial wick assembly.

A7.3 Manufacturing Techniques

1228103   5 May 1969. A Euratom patent describing a method for bon-
(UK)   ding a mesh to a plate, which is then formed into a tube
          for a heat pipe. When the plate is seam welded, any art-
          erial system located along the seam will be retained in
          place.

1125485   27 Jan. 1966. Euratom patent describing a groove manu-
(UK)   facturing technique.

1194530   20 May 1969. Metallgesellschaft Ag. patent concerning ad-
(UK)   ditives to lithium/refractory metal heat pipes as oxygen
          stabilizing agents. Yttrium or other rare elements are
          cited as stabilizing agents.

3620298   22 July 1970. McDonnell Douglas Corporation patent on
(US)   means for joining arteries in heat pipes having right
          angle bends etc. using a connector.

1313525   5 Oct. 1970. Brown Boveri patent describing a process for
(UK)   vapour deposition on heat pipe walls to produce a porous
          layer acting as a wick.

3753364   8 Feb. 1971. Q-Dot patent covering a technique for form-
(US)   ing spirally grooved heat pipes using a special cutting
          tool.

1433542   24 July 1973. Q-Dot follow-up to US Patent 3753364. A
(UK)   related patent is listed in Section A7.4 and the reader
          is also advised to consult UK Patents 1433543 and 1433544,
          published in 1976, and also covering the same topic.

A7.4 Applications

3302042   23 Oct. 1965. Covers application of high temperature
(US)   heat pipes to nuclear reactors, in particular those em-
          ploying thermionic converters.

1183145   29 Jan. 1968. RCA patent on heat pipe cooling of electron
(UK)   tubes.

3607209   26 May 1969. Thermo-Electron patent on use of heat pipes
(US)   to melt glass in furnaces uniformly. Heat is supplied to
           the heat pipes by burners.

1255114   6 May 1970. An air-to-air heat exchanger is described, in
(UK)   which two air ducts are linked by common heat pipes, the
           evaporators being in one duct and the condensers being
           in the duct where the air is required to be heated. (Q-
           Dot Corporation).

3651865   21 Aug. 1970. A heat pipe used to cool electronic com-
(US)   ponents is described. One embodiment employs an inert
           gas for temperature control.

3662542   3 Sept. 1969. An air heater which recovers heat from en-
(US)   gine exhaust ducts is proposed for air conditioning, cab-
           in temperature control etc.

1288222   25 June 1969. A fuse is described using capillary action
(UK)   to carry condensate along the fusible conductor. Once ex-
           cess current is passed, burn-out occurs and the fuse
           fails.

3715610   7 March 1972. General Electric patent on cooling rotating
(US)   machinery using rotating heat pipes.

1304771   23 Jan. 1970. TRW patent on an isothermal furnace muffle.
(UK)

3788389   25 Aug. 1971. This patent describes a structural support
(US)   system to stabilise permafrost. (See Chapter 7).

1433541   24 July 1973. Q-Dot Corporation patent covering heat pipe
           heat recovery units, with particular reference to inter-
           nal structures preventing entrainment. (See also Sec-
           tion A7.3).

A7.5 Variable Conductance Heat Pipes

3525386   22 Jan. 1969. Grover describes a variable conductance
(US)   heat pipe using inert gas control to vary the condenser
           volume.

1222310   30 Dec. 1969. TRW systems techniques for controlling va-
(UK)   pour and liquid flow rates in heat pipes, including valves,
           areas full of wick to allow liquid flow only, and other
           techniques. Suggested applications include space suit
           thermoregulation.

3613773   7 Dec. 1964. (Patent accepted 1971) RCA. This describes
(US)   the conventional VCHP having an inert gas reservoir at
           one end. A cold non-wicked reservoir is proposed.

1238609   25 Nov. 1969. This Euratom patent extends the development
(UK)   of VCHP's to cover control of the temperature of the
           reservoir by locating it in an annulus formed by a sec-
           ond heat pipe directly connected to that at which evap-
           orator control is required.

3958627   15 Oct. 1974. Grumman Aerospace Corporation. A variable
           conductance heat pipe having a baffle which changes the
           direction of vapour flow, permitting a circuitous but
           compact path. As well as being compact, such a unit is
           claimed to have a higher heat transport capability than
           other VCHP's and does not suffer from problems inherent
           in arterial VCHP's (see Chapter 6).

A7.6 Other Types of Heat Pipe

3561525   2nd July 1969. Baer's osmotic heat pipe. (See Chapter 5).
(US)

3563309   16 Sept. 1968. Basiulis describes a heat pipe having an
(US)   improved dielectric strength for heat transfer between
           regions at differing potentials.

3568762   23 May 1967. An RCA application in which a vapour duct
(US)   is located inside the heat pipe to prevent entrainment
           of the liquid in regions of counter-current flow.

3587725   16 Oct. 1968. A uni-directional heat pipe is described
(US)   (one which can transport heat in one direction only).
           This is implemented by having more wick volume at the
           desired evaporator section than in other regions.

3603382   3 Nov. 1969. An annular heat pipe is described which can
(US)   act as a radial heat flux transformer, radial wicking
           being provided to return condensate to the inner annulus.

3613778   3 Mar. 1969. A flat plate heat pipe using porous metal
(US)   layers or mesh layers with vapour channels is described.

1281272   14 April 1970. This Brown Boveri patent covers the use
(UK)   of wetting agents in water heat pipes to aid capillary
           action. One agent suggested is a sodium salt of an alkyl
           naphthalene sulphonic acid ('Nekal'). From examples given,
           the additives only increased capability when wetting was
           not fully achieved by conventional means.

1203332   5 Jan. 1970. Siemens propose a rotating heat pipe (wick-
(UK)   less) for cooling electrical machinery.

3677337   10 Sept. 1970. A further patent on a heat pipe employing
(US)   osmosis.

3700028   10 Dec. 1970. A uni-directional heat pipe in which the
(US)   wick in the preferred condenser section is set away from
             the wall, preventing this area from being used as an ev-
             aporator.

1327794   11 June 1970. Marconi have patented a heat pipe having a
(UK)   comformable wall which can be pressed into close contact
             with electronics components, saving the complexity of
             mounting devices directly onto heat pipes.

1361505   8 Oct. 1971. IRD propose a system in which wicks are used
(UK)   to cover components, these leading to a liquid reservoir,
             cutting out the heat pipe wall interface. The complete
             module is sealed.

3965970   10 Oct. 1974. UK Government. This patent (and its British
(US)   original) covers the inverse thermosyphon as described
             in Chapter 5 and developed by the National Engineering
             Laboratory.

Appendix 8
Conversion Factors

Physical Quantity

Mass	1 lb	=	0.4536 kg
Length	1 ft	=	0.3048 m
	1 in	=	0.0254 m
Area	1 ft^2	=	0.0929 m^2
Force	1 lbf	=	4.448 N
Energy	1 Btu	=	1.055 kJ
	1 kWh	=	3.6 MJ
Power	1 hp	=	745.7 W
Pressure	1 lbf/in^2	=	6894.76 N/m^2
	1 bar	=	10^5 N/m^2
	1 atm.	=	101.325 kN/m^2
	1 torr	=	133.322 N/m^2

Dynamic Viscosity

1 Poise = 0.1 Ns/m^2

Kinematic Viscosity

1 stoke = 10^{-4} m^2/s

Heat Flow

1 Btu/h = 0.2931 W

Heat flux

1 Btu/ft^{2}h = 3.155 W/m^2

Thermal Conductivity

1 Btu/ft^{2}h $^\circ$F/ft = 1.731 W/m^2 $^\circ$C/m

Heat Transfer Coefficient

1 Btu/ft^{2}h $^\circ$F = 5.678 W/m^2 $^\circ$C

Nomenclature

(Many symbols are defined as they arise in the text)

A_c Circumferential flow area

A_w Wick cross sectional area

C_p Specific heat of vapour, constant pressure

C_v Specific heat of vapour, constant volume

D Sphere density in Blake-Kozeny equation

H Constant in the Ramsey-Shields-Eötvös equation

J = 4.18 J/gm mechanical equivalent of heat

K Wick permeability

L Enthalpy of vaporisation or latent heat of vaporisation

M Molecular weight

M Mach Number

M Figure of merit

N Number of grooves or channels

N_u Nusselt Number

P_r Prandtl Number

P Pressure

ΔP Pressure difference

$\Delta P_{c\ max}$ Maximum capillary head

ΔP_ℓ Pressure drop in the liquid

ΔP_v Pressure drop in the vapour

ΔP_g Pressure drop due to gravity

P_{vs} Sink vapour pressure (Ch 6)

P_{va} Active zone vapour pressure (Ch 6)

Q Quantity of heat

R Radius of curvature of liquid surface

R_g Universal gas constant (non-condensable) (Ch 6)

R_o Universal gas constant $= 8.3 \times 10^3$ J/K kg mol

R_e Reynolds Number

R_r Radial Reynolds Number

R_{eb} A bubble Reynolds Number

R_s Heat transfer resistance (Ch 6)

S   Volume flow per second

$S_{1,2}$, S   Control parameters (Ch 6)

T   Absolute temperature

T_c Critical temperature

T_v Vapour temperature

ΔT_s Superheat temperature

T_s Sink temperature (Ch 6)

T_w Heated surface temperature

V   Volume

V_c Volume of condenser

V_R Volume of gas reservoir

W_e Weber Number

a   Groove width

a   Radius of tube

b   Constant in the Hagen-Poiseuille Equation

c   Velocity of sound

d_a Artery diameter

d_w Wire diameter

f   Force

g   Acceleration due to gravity

g_c Rohsenhow correlation

h   Capillary height, artery height, coefficient of heat transfer

k   Constant in Equation 2.3

k   Boltzmanns Constant = 1.38×10^{-23} J/K

k_w Wick thermal conductivity - k_s solid phase, k_ℓ liquid phase

ℓ Length of heat pipe section defined by subscripts, Figure 2a

ℓ Length in Figure 2.1

ℓ_{eff} Effective length of heat pipe

m   Mass

m_g Mass of non-condensable gas (Ch 6)

m   Mass of molecule

$\dot{m}$ Mass flow

n   Number of molecules per unit volume

q   Heat flux

r   Radius

r   Radial co-ordinate

r_e Radius in the evaporator section

r_c Radius in the condensing section

r_H Hydraulic radius

r_v Radius of vapour space

r_w Wick radius

t_p Maximum overshoot/undershoot temperature (Ch 6)

t_r Recovery time (Ch 6)

u   Radial velocity

v   Axial velocity

x   Distance defined in Figure 2.1, and co-ordinate

y   Co-ordinate

z   Co-ordinate

β	Defined as $(1 + k_s/k_\ell)/(1 - k_s/k_\ell)$
δ	Constant in Hsu Formula - thermal layer thickness
ϵ	Fractional voidage. Defined in formula 2.4.2
θ	Contact angle
ϕ	Inclination of heat pipe
ϕ_c	Function of channel aspect ratio (Figure 3.6)
λ	Characteristic dimension of liquid/vapour interface
μ	Viscosity
μ_ℓ	Dynamic viscosity of liquid
μ_v	Dynamic viscosity of vapour
γ	Ratio of specific heats
ψ	Overshoot or undershoot on VCHP (Ch 6)
ρ	Density
ρ_ℓ	Density of liquid
ρ_v	Density of vapour
$\sigma =$	σ_{LV} used for surface energy where there is no ambiguity
σ_{SL}	Surface energy between solid liquid
σ_{LV}	Surface energy between liquid vapour
σ_{SV}	Surface energy between solid vapour
τ	Time constants (Ch 6)

Index